1-10/2015 -7/16
5-1/2017 -3/18

MOVE

A Tale of "O":
On Being Different in an Organization
(with Barry A. Stein)

Life in Organizations:
Workplaces as People Experience Them
(with Barry A. Stein)

Men and Women of the Corporation

Work and Family in the United States:
Critical Review and Agenda for Research and Policy

Another Voice:
Feminist Perspectives on Social Life and Social Science
(coedited with Marcia Millman)

Communes:
Creating and Managing the Collective Life
(editor)

Commitment and Community:
Communes and Utopias in Sociological Perspective

MOVE

PUTTING AMERICA'S
INFRASTRUCTURE
BACK IN THE LEAD

ROSABETH MOSS KANTER

W. W. NORTON & COMPANY

New York | London

For information about permission to reproduce selections from this book,
write to Permissions, W. W. Norton & Company, Inc.,
500 Fifth Avenue, New York, NY 10110

For information about special discounts for bulk purchases, please contact
W. W. Norton Special Sales at specialsales@wwnorton.com or 800-233-4830

Manufacturing by Courier Westford
Book design by Chris Welch
Production manager: Anna Oler

ISBN: 978-0-393-24680-3

W. W. Norton & Company, Inc.
500 Fifth Avenue, New York, N.Y. 10110
www.wwnorton.com

W. W. Norton & Company Ltd.
Castle House, 75/76 Wells Street, London W1T 3QT

1 2 3 4 5 6 7 8 9 0

TO ALISON LILY STEIN, NATALIE ELYSE STEIN, AND
JACOB HARRISON STEIN, WHOSE FAVORITE STORYBOOKS,
INCLUDING *THE LITTLE ENGINE THAT COULD* AND DR. SEUSS'S
OH, THE PLACES YOU'LL GO!, REMIND US THAT
DETERMINATION AND CREATIVITY CAN ALWAYS FREE US
FROM GETTING STUCK IN "WAITING PLACES" AND HELP US
MOVE TO A BRIGHTER FUTURE

CONTENTS

PREFACE: IN SEARCH OF LEADERSHIP   1

1. STUCK ON THE WAY TO THE FUTURE   9

2. ON THE RAILS   27

3. UP IN THE AIR   73

4. SMART ROADS MEET THE SMARTPHONE   119

5. RETHINKING CITIES   171

6. THE WILL AND THE WALLET   221

7. HOW TO MOVE   259

NOTES   281

ACKNOWLEDGMENTS   301

INDEX   305

IN SEARCH OF LEADERSHIP

I was planning to write a book about leadership. Then the infrastructure issue struck me and took me on a detour.

It struck me in the head when I saw data about dissatisfaction with the American transportation system—congested roads, crumbling bridges, middling passenger railroads, and air traffic delays. It struck me in the foot when I tripped on a pit at the edge of a bridge repair site, causing me to limp for days and to take a car to work instead of walking. It struck me in the lungs while breathing fume-filled air on a stop-and-go drive on a so-called expressway, giving me time to contemplate health implications and global warming. It struck me in the heart when I learned of inner-city transit problems at a meeting of a nonprofit organization I serve, such as hours-long commutes by multiple buses just to go a few miles to get an education. It struck my patriotic pride when I rode the Shanghai subway and maglev train, marveling at speed and efficiency unknown in the United States, after being equally impressed by the ultramodern airport. It struck my imagination when I contemplated visionary innovations in transportation on the horizon, such as automated cars or smartphone apps promising to make driving safer or parking faster.

In short, the topic started out as an intellectual abstraction but soon became personal. Since the problems strike us every day, why is it so hard to make progress? And what should be done?

I thought: this is an important topic, even if it requires putting the leadership book on hold. I felt that the problem needed a voice coming from the people—from consumers, families, business executives, entrepreneurs, community leaders, and the public. There are plenty of experts analyzing the technical issues and doing it well, whether civil engineers, economists, or public policy specialists. What is needed is a human face and human interest. I decided to look at the topic from the perspective of users.

A pivotal conversation in 2012 showed that there is a great deal of energy for change. I spoke with former U.S. Secretary of Transportation Rodney Slater, who served in President Bill Clinton's cabinet and was one of the fellows in the Advanced Leadership Initiative that I cofounded and direct at Harvard—a new stage of higher education for leaders transitioning to their next years of service, using every field and discipline at the university. As the result of work in the late 1990s, the Clinton administration had issued a twenty-five-year transportation vision, but subsequent events put this on hold. Secretary Slater and his peers felt that the public still needs to be rallied behind a transportation vision for the twenty-first century.

Inspired by the potential, I committed to examine transportation and infrastructure as part of the Harvard Business School U.S. Competitiveness Project, an effort by my globally minded institution to ensure that America continues to lead rather than fall behind. I mounted an extensive twenty-month effort to understand a multifaceted system and find the best representative examples of pain points and bottlenecks, as well as promising models for the future. I mobilized a research team of full-time professionals. We reviewed nearly one hundred reports from as many associations and think tanks, compiled statistics (and tried to get behind numbers that sometimes conflicted), drew on faculty experts, went on behind-the-scenes field visits, conducted hundreds of interviews, and in February 2014 convened a national summit at HBS, which I chaired. The summit included nearly two hundred industry and government leaders, airline and railroad CEOs, technology innovators, labor leaders, national association heads, community activists, and public officials, including mayors, U.S. senators, and members of Congress—leaders across sectors and industries who had rarely sat in the same room with one another. One conclusion was that the wider public needed to be part of the discussion. That encouraged me to contemplate writing a book that could engage a broad audience.

I went on the road and took to the air. Throughout the United States, I listened to numerous innovators, such as Zipcar's former CEO Scott Griffith and the founder of a venture producing flying cars, and investors seeking promising projects as they saw the potential for infrastructure and transportation investments. Firsthand investigations included visits to a tunnel under the ocean in Miami in its final construction phase; the streets, bike-sharing stations, and rail crossings of Chicago, guided by local planners and officials; the American Airlines flight operations center in Texas, meeting with pilots, dispatchers, and experts from the Weather Company; the Verizon Innovation Center, with its new applications of technology under development; and the latest instant bridges, among other tours. I had frequent conversations with a railroad CEO at board meetings we shared and with airline executives I knew. My team and I went to downtown offices and run-down inner-city neighborhoods, riding various forms of transit, while conducting interviews with dozens of people in half a dozen cities.

Given that infrastructure often seems to be a man's world, even though women are major consumers and elected officials, it was heartening to include outstanding woman leaders in the interviews and visits, such as former FAA Administrator Jane Garvey, serial public transportation chief Beverly Scott, logistics company CEO Ann Drake, MIT robotics lab head Daniela Rus, General Motors CEO Mary Barra, and IBM CEO Virginia Rometty. Barra and Rometty attended college at General Motors Institute around the same time, proving that girls like cars, too, and that they can become engineers.

The effort was national in scope but necessarily demanded focus. A few places received comprehensive attention: the Chicago metropolitan area, Miami and Miami-Dade County, and various parts of

California, New York, and Massachusetts. Massachusetts appears often in this book not because Boston is my home base but because the Commonwealth has been a transportation pioneer and hotbed of innovation. We also viewed Atlanta, Detroit, Philadelphia, Denver, and other metropolitan regions from both ground and air. For a comparative perspective, I drew on my international travels to China, Singapore, Japan, and Europe for informal observations and conversations. For example, I met the CEO of BMW in Munich and the CEO of Turkish Airlines in Istanbul. Earlier, I had welcomed the former mayor of Curitiba in Brazil, an urban transportation pioneer, to a think tank for my Advanced Leadership program, and I had met other mayors and teams from global cities at IBM's Smarter Cities Challenge Summit. As the research progressed, I found that many of the ideas I was developing were in sync with enlightened officials of local, state, and federal departments of transportation who are passionate about improvements. I spoke with a range of other elected officials while presenting at a program for newly elected mayors at Harvard Kennedy School of Government.

Some analysis could build on my previous work, which centers on strategy, innovation, and leadership for change. My book *The Change Masters* included a long examination of General Motors and the auto industry during an earlier crisis, before the recent government bailout and safety issues. *Confidence: How Winning Streaks and Losing Streaks Begin and End* featured the turnaround of an airline against the backdrop of industry challenges. *World Class: Thriving Locally in the Global Economy* focused on strategies for revitalizing cities, including airports and ports as foundational local assets. I have written extensively about innovation and know many technology companies from the inside, as a consultant and adviser. I have served on government councils and commissions for which infrastructure and transportation are key considerations. All of this proved relevant experience for dissecting a very big and messy change problem.

Writing about a sector of current importance is challenging, because situations are constantly changing. Some aspect of transportation and infrastructure is in the news daily. Problems surface, crises ensue, blame is assigned, projects are announced, political debates take place, elected officials and CEOs leave office, all while businesses merge and acquire, and start-ups come and go. It's a delicate balance to connect to current events while presenting a larger, longer-lasting picture. A book can offer the big picture, not just the elements but the whole, tying together many otherwise isolated events and parts of the system. The ideas and stories can provide guidance to interpret what's happening (or not happening)—a strategic vision rather than single actions. This is vital if we want to move beyond short-term fixes to finding enduring solutions. If the focus is only on the crisis du jour, then new problems will constantly pop up.

I take the issues personally, as I said, and people I meet at professional and social events do too—maybe too personally. When I mention that I'm working on a book about transportation and infrastructure, many people want to tell me their own tales of woe. Their potholes. Their long commutes. Their awful flights. Everyone has a story, followed by a whine: "Why don't they do something about it?" (Congress, the president, the airlines). After listening carefully, I began to see another challenge: how to make the story about us, not them. Our roads. Our commutes. Our flights. Our environment. Another goal of this book is to provide one picture of our shared situation. Unless we see a common fate, it is hard to get consensus for action.

One frequent question in these conversations is why other countries can have world-class transportation systems, and we can't. I respond that it's not because they're smarter, richer, or smaller, and thus potentially more manageable. It's a matter of priorities and politics, as will become abundantly clear throughout this book. America has the expertise and innovation, and there's money to be found, as I show later. There is an emerging consensus about the importance of buses and other forms of public transit. But nations that are outperforming the United States in these areas have greater faith in government and allocate public money for public works at the national level.

In comparison, the U.S. has coasted on past successes, failed to fully confront mounting problems, lacked sufficient vision about future opportunities, and, in Congress, held essential funding hostage to partisan battles over taxes. Although federal stimulus funds were authorized for shovel-ready projects following the financial crisis, these fell far short of the amounts needed for either infrastructure repairs or fast economic recovery. High-speed rail has been a priority of several recent U.S. presidents, yet has failed to gain traction, although California has ambitious plans. Despite the certainty that red states' bridges can crumble as readily as blue states', congressional representatives elected every two years can't be expected to rally behind raising taxes to pay for big projects, unless there's strong public will that translates into votes. Indeed, one solution to the politics is to depoliticize infrastructure decisions, as other nations have done by way of infrastructure banks, which create dedicated pools of long-term capital not dependent on taxes or on yearly appropriations of funds—a concept thus far dead in the water even in the relatively longer-term-thinking Senate. But this isn't a book about Congress.

Albert Einstein's famous definition of insanity is doing the same thing over and over again and expecting a different result. Continuing to point fingers at a dysfunctional Congress or to rail against inefficient bureaucracies won't get America moving. It's time to rethink how the problem is approached. It's time to step back to identify the roots of current problems and find new sources of solutions, especially solutions that build on U.S. strengths in technology innovation and civic coalitions. It's time to look with fresh eyes and write a new story.

So this isn't a Washington-centered book; it's a book about the many places that are taking action. It doesn't get into tired debates about the role of the federal government versus that of the states, or whether there should be public spending at all. Of course, there should be. Examining transportation infrastructure from the perspective of the local and regional user—for instance, the potholes or traffic delays on all of our streets—makes clear that national funding is essential; states and cities can't possibly raise enough money themselves for all the work that must be done, and transportation doesn't stop at municipal or state borders anyway.

One final question motivated me: if not Washington alone, then where and who?

Cross-sector innovation and collaboration are big parts of the solution. Laboratories of democracy at the state and the local levels can mount tests; as I write, Oregon is running an experiment with vehicles-miles-traveled fees to replace gasoline taxes for roads and bridges, which assumes the spread of electric vehicles that use roads but don't consume gas. Private-sector innovation, ranging from experimental driverless cars from Google's transportation division to proliferating smartphone apps for finding parking or renting out your driveway, combines with public-sector agencies, such as Ford's and GM's work with the Michigan Department of Transportation on automated vehicles, or Boston's work with tech companies on its own parking app. Collaboration in the form of public-private partnerships can potentially tap diverse forms of funding, combine expertise, and make it possible to get public buy-in. Federal funds can come in many forms, not just outright grants: loans, loan guarantees, seed money.

We can sound the alarm and then shift the paradigm to give people reason for hope. Certainly the future trends identified here—revitalized cities with people-friendly streets, smarter roads and vehicles that can prevent accidents, new technology to reduce weather-related flight delays, public transit improvements—are causes for optimism—if we can support such changes. They also present significant business opportunities.

For all the concerns and problems, this is ultimately a hopeful book. The facts, figures, and stories about problems and opportunities are inherently interesting, and they can serve as a call to action. Fixing infrastructure is a way to rethink the nation. Infrastructure is important in itself but is also a good starting point for generating the courage and collaboration that will help us get moving to solve a range of other problems. We need inspiring visions, strategic thinking, openness to innovation, and change processes that involve coalition building and uniting constituencies behind common goals—the essence of leadership. We need consumers, citizens, activists, corporate chiefs, and enlightened officials who are informed and motivated to seek change.

Now I see that my detour was short and that it has brought me full circle. I'm still writing about leadership. It's essential to understanding how to build a strong American future.

1

STUCK ON THE WAY TO THE FUTURE

G oing somewhere? Not so fast. Expect delays ahead.

- The average American commuter wastes a total of 38 hours in traffic per year. This amounts to 5.5 billion hours in lost U.S. productivity annually and 2.9 billion gallons of wasted fuel. For 2011, the cost of congestion in wasted time and fuel was estimated to total $121 billion, or $818 per auto commuter.[1] Traffic congestion alone costs about $70 billion per year in time wasted, and the public health cost of pollution from that congestion is about $15 billion per year, according to Harvard School of Public Health researchers.[2]

- About 30,000 people who normally commuted by train had to take to their cars after a 2013 commuter train crash in Bridgeport, Connecticut, which injured seventy-six, cost $18 million in economic damage, and closed all eastbound rail traffic from New York all the way to Boston. This clogged the roads between New York and New Haven for the week it took to fix just 2,000 feet of track, which should have been done in a day. The cause of the crash was traced to a broken rail joint that had been due for maintenance for a long time, and was surrounded by components such as electric wiring dating to the early 1900s.[3] The disruption created a massive mess of delayed commuters, missed appointments, obstructed emergency vehicles, and air pollution skyrocketing from the idling cars. This added to already record levels of traffic congestion in the Northeast Corridor, America's busiest, whether on the roads, on the rails, or in the air, and revealed other delay-causing infrastructure problems. A corporate CEO claimed that he counted at least 1,278 potholes as he drove between New York and Boston; it's hard to believe either the counting or the number, but clearly there are a lot of potholes.

- Between 1989 and 2013, the United States had nearly 600 bridge failures. A bridge collapse in Minneapolis in 2007 killed thirteen people and injured over a hundred others.[4] In 2012, a quarter of all U.S. bridges were deemed by the Federal Highway Administration to be structurally deficient or functionally obsolete. In 2013, those numbers dropped only very slightly: 24.3 percent are structurally deficient or functionally obsolete—a whopping total of almost 64,000 bridges.[5] Bridges are typically designed for a fifty-year life span; by 2023, one in four U.S. bridges will be over sixty-five years old, the average age of structurally deficient bridges today.[6] Fully 42 percent of the bridges in Rhode Island could be structurally deficient within seven years, says Director of Transportation Michael Lewis.

- Delayed or canceled flights cost the economy an estimated $30 billion–$40 billion a year.[7] In the severe-weather month of January 2014, thousands of flights were disrupted. Ice storms in Atlanta, predicted in advance but not prepared for, forced some people to stay in their cars for as much as twenty or more hours without moving.

- A single mother in inner-city Chicago gets up early to take her children on a bus to a daycare center; then she boards three additional buses in order to get to her suburban job, about eight miles from her home. The trip takes her some two and a half hours each way. She can barely make the 7 p.m. closing of the child care center. She can't afford a car, and there are no direct public transportation options in her neighborhood. She is one of many people around the country waiting for better options, only to see funds designated for public transit diverted to pay for roads or public debt instead.

Whether daily commuter, frequent flier, employer or associate, entrepreneur, part-time worker, schoolkid, college student, community leader, suburban shopper, e-commerce customer, or inner-city job seeker, everyone is affected by the transportation system and the platforms it runs on. If people can't move, if goods are delayed, and if information networks can't connect, then economic opportunity and quality of life deteriorate. To move is to thrive. To be stuck is to lack opportunity. Once the Land of Opportunity, America seems to be entering the State of Delays. There are traffic delays, travel delays, shipping delays, repair delays, project approval delays, delays caused by underfunding, and delays in revisiting and revising obsolete assumptions.

For sharp contrasts, take a quick trip around the world. Whiz through the "Chunnel" connecting England and France, ride a high-speed train in China or the maglev from Shanghai Pudong Airport to the city, get high-speed Internet and cell phone connections on a remote mountain in Turkey, hop a constantly circulating speedy bus-on-a-platform in Curitiba, Brazil, or travel in a driverless Mercedes in Germany, and see a future that the United States is barely glimpsing.

Of course, America continues to enjoy considerable strengths. International comparisons show that there are many things the United States does extremely well, which are models for the world. Most of them fall into the realm of entrepreneurship and innovation, aided by a leading system of higher education and a culture that values start-ups and private enterprise. The United States gave birth to smartphones, apps, and social media. U.S. information technology companies are global leaders. IBM introduced the idea of a "smarter planet" (meaning information enabled) and partners with Google in cloud computing and more recently with Apple for iPads running IBM enterprise software.

But the deployment of our own technology has occasionally lagged. I asked IBM CEO Ginni Rometty about smarter transportation and infrastructure over a quiet lunch. She identified exciting projects that use data analytics to deal with traffic congestion and fuel management, but they were almost entirely outside the United States. Other nations have also been faster to tackle big projects requiring public sector involvement. In the time it takes United States entrepreneurs to create parking apps one city at a time, other nations might revamp an entire national system. The United States has world-leading telecommunications companies but not yet full cell phone coverage everywhere. General Electric CEO Jeffrey Immelt attests to that, having almost lost a deal as a result of spotty cell phone services in Connecticut. While on the phone talking with a top executive in Nigeria, Immelt's cell service nearly cut out; he had to ask the African to wait while he reached a landline to call him back. (Hmm. Which is the Third World country?)

The United States is better at entrepreneurship than at joint action that requires public leadership to forge consensus. That's a sweeping generalization, but it helps explain the state of delays. The issues in America are institutional problems that have grown for years and have had disturbing consequences, on both the economic and the human sides.

Consider safety and health. The National Highway Traffic Safety Administration estimated the cost of traffic accidents alone at $871 billion per year,[8] using 2010 data in a report issued in 2014. This estimate combines the harm from loss of life, pain, and decreased quality of life with the economic costs of productivity loss, property damage, medical/rehab services, congestion, legal/court proceedings, emergency services, and insurance administration. In 2010, 13.6 million traffic accidents led to 24 million damaged vehicles, 3.9 million nonfatal injuries, and 32,999 deaths. Fortunately, seatbelts prevented 12,500 deaths and 308,000 serious injuries, saving the economy $69 billion.[9] Imagine what could be done with more-advanced technologies, such as sensors that keep cars in their lanes.

Or take the environment, another area significantly impacted by transportation and infrastructure. The United States generates the largest amount of carbon emissions in the world; it emitted a total of more than 5 billion metric tons in 2012. The transportation sector accounts for 1.7 billion metric tons, or 34 percent of U.S. carbon emissions.[10] In total, the transportation sector accounted for 28 percent of U.S. greenhouse gas emissions, second only to electricity generation in environmental impact. Personal automotive transportation—cars and light trucks, SUVs, and minivans—accounted for an astonishing 61.5 percent of the sector's emissions, or 17 percent of total U.S. carbon emissions. Freight trucks accounted for 21.9 percent of transportation-related emissions, commercial aircraft for 6.2 percent, rail for 2.5 percent, and ships and boats for just 2.2 percent.[11] Traffic congestion has been estimated to contribute 56 billion pounds of excess CO_2 to the atmosphere every year.[12] Imagine the benefits from a rebalancing of transportation policies that increases use of rail or bicycles, or from applying communications technologies in the air that reduce the burning of fuel.

Reinventing transportation and infrastructure has the potential to save lives, cut costs, add convenience by easing congestion, reduce pollution and mitigate climate change, and create future growth opportunities that generate new jobs. It's not just win-win; it's win-win-win-win-win. Seeking these quintuple wins should be the goal of investments in the future, which I emphasize throughout this book. But before tackling the future, we should understand the institutional legacy that got us here.

HOW AMERICA GOT STUCK

I want to name a horror movie *The Street That Devours SUVs*, in honor of a Chicago street suffering from aging water pipes leaky enough to open a hole that swallowed a minivan in 2011. Many relics like those pipes exacerbate current infrastructure problems, and they're not the kind of relic that produces nostalgia or that we'd want to reintroduce, like vinyl records. Some infrastructure antiques can be replaced, such as 100-year-old overhead wiring. More challenging are legacy structures that produce patterns of use, such as streets laid out when streetcars reigned, or city limits defined by century-old rail lines. But even those seem small compared with one gargantuan effect.

The giant from the past that has America stuck in history and stuck in traffic is the high-speed highway. In the 1950s, starting just after victory in World War II and continuing into the 1960s, the United States constructed new models and crafted policies that have shaped American lives for sixty years. The Interstate Highway System, funded in the 1950s in the name of national defense, made the car the preferred mode of ground transportation. It changed residential patterns, enlarging suburbs and creating exurbs, and made long commutes to work by car the national norm, shoving rail into the background. The legislation that created interstate highways was not named the Drive Out Trains and Bicycles Act, or the Grow the Auto Industry Act, but it did both those things.

In 1956, with President Dwight Eisenhower's strong support on defense grounds, Congress authorized the new national highway system by passing the Federal-Aid Highway Act, commonly known as the National Interstate and Defense Highways Act.[13] The act authorized a then unprecedented $25 billion to build 41,000 miles of superhighways over ten years. It also established the gasoline tax–fed, pay-as-you-go Highway Trust Fund, which remains the primary highway funding vehicle to this day.[14] With collaboration between state departments of transportation (known as DOTs) and the U.S. Department of Defense, planners laid out a highway system that would serve both peacetime and strategic defense needs.[15] Parallel defense-related actions boosted air transportation and funded airports by similar taxes. The Cold War and *Sputnik* provided one more defense-related investment push.

America was at the top of the game in that era. It wasn't all idyllic for everyone, and rumbles of the coming push for civil rights and racial justice were heard. But the moment of peak glory can be the time that decline sets in. It's a rule of nature. And when you're on top of the world—strong, admired, technologically advanced—it's easy to become complacent. Having made enormous investments, you can readily sit back and neglect repairing or updating them. And that's what the United States did. Certainly, new incremental improvements have emerged here and there, especially in telecommunications, aviation technology, and the birth of the Internet. But after the big burst of investment half a century ago, big infrastructure items began to slip and fall behind those of other nations, including some that were poor and war-devastated in the 1950s but are now world leaders. And in the United States, ideas of the 1950s turned into social institutions with taken-for-granted assumptions that began to seem normal and inevitable—so that people would build in extra hours to anticipate traffic delays, but not join a movement to address their root causes.

The transportation sector is large in the United States, but large doesn't mean efficient; it means hard to change. Nearly 20 million Americans work in transportation, transportation infrastructure, and related industries. That number, obtained by adding up different sources of data, includes 1.7 million workers in manufacturing for vehicles, components, or related infrastructure, and nearly half a million in construction for highways, roads, bridges, and heavy infrastructure. Transportation also looms large in individual spending. The average household spends 19 percent of its budget on getting around, according to the Federal Highway Administration.[16] Another analysis reduced the figure to closer to 11 percent for a family with two parents and two children.[17] "When people today in America are spending more money on transportation than they're spending on food and health combined, that's not a good economic system," declares former Miami mayor Manny Diaz.

America is stuck on the cusp of change. Powerful forces of change are sweeping through the world. For the United States, the old era is crumbling—literally, as in the case of bridges. Old assumptions no

longer hold, but the policies endure. The defense rationale for highways sixty years ago is untenable; it should be clear by now that cities
can't be evacuated in an emergency by getting people to pile into their
cars and drive out, as was shown tragically after Hurricane Katrina
hit New Orleans. Six decades after roads became a priority, the nation
has changed. Central cities have revitalized and made a residential
comeback. Traffic congestion and pollution are recognized as significant problems, contributing to unsustainable use of resources and to
climate change. Young professionals, and many others, are no longer
fully in love with cars as a status symbol or with joy rides as entertainment; they want alternatives, such as good public transportation.

Smartphones and wireless networks have come into prominence
fast and provoked change. Communications networks have emerged
as a major type of infrastructure in and of themselves, serving as an
innovation platform and as drivers of a revolution in how we move
people and goods. Already emergent possibilities include the use of
technology to empower individual choices, ranging from on-demand
transportation to pilots' communicating in real time with air traffic
controllers to make adjustments in the air. A millennial generation
brings environmental values and a move toward a "sharing economy" of "collaborative consumption," instead of individual ownership. Visions of pedestrian- and bicycle-friendly cities, with greener/
cleaner environments, are already being realized in many places.

Attitudes are changing for those at the top of the income distribution as well as for ordinary people. Nearly two thousand business
leaders responding to the third annual Harvard Business School
U.S. competitiveness survey in December 2013 echoed familiar
complaints about problems with traffic congestion, airports, roads,
and transportation for their employees. Most respondents said that
they had seen no changes or improvements over the past three years,
but of those who did notice changes, up to a five-to-one majority felt
things were getting worse in most respects. And the single item getting the highest degree of consensus as a solution, identified by 40
percent of the respondents, is the need for more and better public

transportation to more places. That is a high proportion and reflects a concern for others rather than for themselves, since this group can easily find private solutions, whether limousines or NetJets.

In combination, the forces for change challenge conventional policies and approaches. A gasoline tax to fund highway repairs seems a quaint creature of the past. It was already quaint even in the 1970s when a small portion was opened for use for mass transit. In the summer of 2014, the Highway Trust Fund almost ran out of money because gas taxes just don't generate the revenues they did in gas-guzzling days. Fuel standards have made cars more fuel efficient, electric cars don't use gas at all, and car ownership is declining, especially among urban residents who prefer alternative means of transportation. We're between paradigms. While gas taxes wane— although states are raising them—there's no consensus about what will replace or supplement them. Will it be user fees set by electronic tolling, with transponders and video cameras, or annual vehicle-miles-traveled fees, or something else?

Taken-for-granted assumptions must be revisited to make way for a different future, one in which technology plays a greater role, and infrastructure is broadened to include information and communications networks. The digital age now enables some people to stay home and work remotely, hold meetings by Skype, get education from online courses, find entertainment from streaming video, use a ride-sharing app for occasional trips beyond the neighborhood, and order groceries and other goods online or, perhaps in the near future, print items at home on 3-D printers. Physical infrastructure is still important, such as cell towers and delivery trucks. But the emphasis is shifting from fixed to dynamic infrastructure, from highways to the Information Superhighway.

The continuum of action ranges along three R's: repair, renewal, and reinvention. It's clear that all three R's must be applied to the problems, with a tilt toward reinvention in order to tackle the potential opportunities. The need for change is urgent, but where is the sense of urgency?

CAN WE MOVE?

In 2000, the U.S. Department of Transportation issued a road map for the next twenty-five years, the first long-term plan since President Gerald Ford's Secretary of Transportation, William Coleman, was in office.[18] The plan followed notable successes guided by President Bill Clinton's Transportation Secretary Rodney Slater, which included a then-record $200 billion in funding for surface transportation and forty Open Skies agreements with other nations.

Then, in 2001, events intervened. The economy turned sour with the dot-com crash even before the terrorist attacks of September 11. Other spending priorities soon overwhelmed the federal budget. For the opening decade of the twenty-first century, national leaders turned their attention to other matters, including a global financial crisis and the Great Recession. Aging infrastructure became the focus of an economic stimulus program to fund shovel-ready projects, often road repairs, that would create construction jobs.

On February 26, 2014, in a speech in St. Paul, Minnesota, President Barack Obama announced that he was seeking $302 billion in funds for transportation and infrastructure. About two hours later, I opened a national summit at Harvard Business School, "America on the Move: Transportation and Infrastructure for the 21st Century." There was overwhelming consensus among the participating business, labor, government, association, and community leaders about the importance of the issues and overwhelming concern that America was not moving fast enough to do enough. A gridlocked, overly partisan Congress at odds with the president is often blamed for delays in national action, and the lawmakers deserve it. But even partisans respond to voters. The public needs to be mobilized to demand change, and fast. We must find common purpose.

Pick your issue, and you'll find a transportation infrastructure angle. *Business*: The state of transportation and infrastructure affects costs, the speed with which products can get to customers, and the location of production facilities, warehouses, or retail outlets. And, of course, companies directly in the industry care about infrastructure to support their mode of transportation: air, rail, auto, or whatever else. *Employment*: Traveling to work is the major use of highways and mass transit, but public transit is not always available or convenient. *Regional economic development*: Opening new trade or tourism routes, being ready for new opportunities when shipping patterns change, upgrading neighborhoods, and developing new areas to be incubators that attract entrepreneurs all depend on transportation and infrastructure to ensure everything from long-distance connections to parking availability. *Family*: Transportation and infrastructure affect the safety of adults and children, the distance to work, commuting times, and access to services, such as shopping, health care, or recreation. *Education*: How kids get to school and how long it takes influence where and how well they can learn, and school systems are major operators of buses. *Health*: Poor transportation and infrastructure and associated accidents can cause injuries and even death, affect health system costs, and, by limiting opportunities for walking or biking, contribute to an obesity epidemic. *Environment*: The air we breathe and greenhouse gases in the atmosphere are influenced almost more by current transportation modes than by anything else. *Social justice*: The poor are often stuck in urban neighborhoods with the fewest transportation options for getting to jobs, child care, health care, and discount big-box stores; that makes it harder for them to climb out of poverty and exacerbates inequality.

Although everyone cares about those issues in some broad sense, responsibility for them is diffused, and people see only the portion that looms largest to them.

The problem of transportation and infrastructure is divided up into silos within the government and into silos across industries. Hundreds of major associations, coalitions, and think tanks advocate for their own mode of transportation, issue by issue—and then coalitions of coalitions try to link them. Multiple levels of government are involved—federal, state, local, and regional authorities—and within them, departmental silos separate, for example, transportation, communications, environmental issues, economic development, and technology. Across industries, air and rail don't talk to each other (although freight rail and trucking companies work well together). Furthermore, the major investors in U.S. infrastructure are often foreign companies; a French firm financed the Miami port tunnel, paying a French construction company that used a German tunnel-boring machine, and another French company won the contract to operate Massachusetts Bay Transportation Authority subways and trains. There's nothing inherently wrong with taking foreign capital or using foreign expertise, but unless U.S. businesses have a bigger stake in the issue, they can't be knowledgeable contributors to public-private planning. America has gotten stuck because connections across relevant players are lacking. Transportation is all about connecting one thing to another. It will take new connections to get us moving.

Forces for change are clearly not matched by the urgency of action. Given so many situations and factors that should arouse enormous concern, why is it so hard to secure public support for long-term infrastructure investments and to get Congress to vote for them? I think it's a structural issue. Silos, narrow interests, and fragmentation mute outrage.

Perhaps we're stuck not only with aging infrastructure but also with obsolete ways of talking about it.

MAINTENANCE IS NOT A VISION: WHAT'S WRONG WITH THE INFRASTRUCTURE DEBATE IN AMERICA

For starters, the word "infrastructure" doesn't sizzle. It sounds technical, inanimate, and bureaucratic. The case for infrastructure is often made by statistical abstractions, not by emphasizing the daily needs of ordinary Americans: how we get to work, find affordable goods and services, take our children to school, or access health care. Transportation touches everything; physical and geographic mobility increase choices and opens opportunity. But there's no consumer movement for infrastructure. Although public transit has numerous grassroots advocacy groups, they rarely if ever use the word "infrastructure" when pressing for better bus service, for example. Long-term investment requires a human face and clarity about benefits. Moreover, it's hard to have a public agenda without a sense of shared public purpose.

America needs a vision that can inspire, unite, and motivate action. The scare factor doesn't rally support, although it can make headlines. Predictions that bridges might crumble and kill people get no attention until they come true, as they did in Minneapolis in 2007. But memories fade, and maintenance leaves you just where you were before, only with a smaller checkbook and bad memories. Almost every day I walk to work across the Charles River via a picturesque, historic, red brick bridge. After several years of stressful reconstruction, there will be (drum roll, please) a picturesque, historic, red brick bridge. It will be the same as before, with no enhancements or technology to ensure peaceful coexistence of pedestrians, buses, cars, trucks, and bicycles.

Maintenance is not a vision. It is often not a priority. Consider a repair-avoiding couple I know in Washington, D.C., who are serial home buyers. When they live in a house long enough that it needs maintenance, they sell it and buy another. Do they epitomize Washington attitudes? It's easier to abandon the old than pay to renew it. That's one reason cities such as Detroit became hollowed-out sites of boarded-up or burned-down neighborhoods as industries declined, but also how poor inner-city areas in every major city have been left with the consequences of obsolete infrastructure.

Immediate job creation is a very important benefit of infrastructure renewal, especially to the workers who get the jobs. Job creation is one theme on which business and labor agree; U.S. Chamber of Commerce president Thomas Donohue and AFL-CIO president Richard Trumka have traveled the nation and Capitol Hill for almost five years making joint presentations on infrastructure as a job-creation mechanism. By itself, however, job creation is not a vision. The nation's big infrastructure spurts, as we shall see in this book, occurred thanks to a grand vision of building an economy not just for the present but for the future as well. The transcontinental railroad, the Interstate Highway System, and the post-*Sputnik* space race were not about jobs. In fact, in the nineteenth century, the railroads imported foreign labor from China to get the tracks laid. Those visionary projects propelled growth and opportunities for a new economy, as remote areas gained transportation links and as new aerospace technologies emerged. Transportation connections, whether public transit stops in a neighborhood or international airline routes, open new possibilities for commerce and trade, which bring multipliers beyond the few construction jobs they entail. Transportation patterns build communities and shape society.

Infrastructure spending sounds big and expensive, at a time when the nation doesn't like either, although private investors are poised to join promising projects that have a clear strategy. But for the same reason, infrastructure might not seem cool to the young. Rising entrepreneurs think small, private, and disruptive, seeking new technologies that bypass establishments. Young urban professionals increasingly don't own cars, a development that reduces their interest in debates about the Highway Trust Fund. Yet, they are busily creating the infrastructure future. Innovators are accelerating development of autonomous vehicles, changing the role of human operators (and the meaning of driver's licenses). Parking apps abound; connected cars are the new wave. Zipcar and Uber use technology for car sharing and ride sharing. A Boston entrepreneur a year out of college, borrowing a page from Google's (controversial) private bus service for employees between Silicon Valley and San Francisco, is creating a new on-demand, data-driven, private urban bus service. To realize these transportation dreams, information-enriched infrastructure is essential. Entrepreneurial techies are rarely at the table when infrastructure is discussed, but they should be. And then perhaps they'd learn to cope with the politics, when local bodies resist their innovations—for example, city councils protecting the taxi industry against Uber.

It shouldn't be as hard as it is to get moving on a shared transportation and infrastructure agenda. Infrastructure has no ideology. Bridges either stay up or fall down. General Electric CEO Jeffrey Immelt, who headed the President's Council on Jobs and Competitiveness after the financial crisis, put infrastructure on the agenda as a central driver of a competitive economy, along with training and education, small businesses and entrepreneurship, and regulatory reform. Governor Deval Patrick of Massachusetts made it one of three priorities for his eight years in office, along with education and jobs.

BRIGHT SPOTS

The magnitude of the problem and the roadblocks to action can seem daunting. But there are numerous bright spots, highlighted throughout this book, in every mode of transportation, that serve as role models and guides to the future.

Leaders have emerged from across the spectrum. Freight rail leader Michael Ward of CSX has guided the company to become greener, winning top environmental awards, and has invested the company's own money in repairing and maintaining tracks. Entrepreneurs like Robin Chase and Antje Danielson and their successor Scott Griffith built Zipcar into an affordable alternative to owning a personal car. Beverly Scott, general manager of the Massachusetts Bay Transportation Authority, is a public transit advocate who wants to serve the underserved and knows how to build community coalitions. Tech-savvy pilot Brian Will of American Airlines is playing a leading role in the movement toward navigation technology that can enable landings by gliding, thus reducing carbon emissions. Jose Abreu, former Florida Secretary of Transportation under Governor Jeb Bush, is a public-sector visionary with long-term persistence; he nurtured a dream of a new tunnel under the Port of Miami for thirty years, even when not in office, and applauded its opening in 2014, accompanied by an immediate reduction in truck traffic in downtown Miami.

For political leadership, look to governors and look locally. With a gridlocked Congress, mayors have become heroes, starting or supporting transportation and infrastructure projects with long time horizons that build on predecessors' actions and extend beyond their terms in office: Michael Bloomberg of New York City, Manny Diaz of the City of Miami, or Rahm Emanuel of Chicago. Bright spots are springing up regionally. Atlanta and Detroit have plans under way for airport–passenger rail connections. Los Angeles is at the center of ambitious plans for urban light rail, airport and seaport connections, and longer-distance high-speed rail, which Governor Jerry Brown has championed for the whole state at a time when high-speed rail is stalled nationally. Chicago has ambitious plans to ease rail congestion, rethink city streets, upgrade the bus system, and increase bicycle-friendly roads with bike-sharing systems. Technology is a powerful tool. Dynamic streets can change pricing to encourage less congestion; vehicle-to-vehicle communications can reduce traffic accidents; data analytics for better weather forecasts can reduce fuel burn and weather-related air delays.

Another bright spot is a growing awareness of the need for regulatory reform. The long, drawn-out regulatory process that has delayed promising projects can be speeded up when the will is there. The U.S. Department of Transportation has moved approvals for New York's new Tappan Zee Bridge at record speed, although other agencies have slowed it down. The Federal Aviation Authority has deployed new technology to help reduce air traffic delays faster than skeptics were ready to believe. Public-private coalitions can set priorities, find synergies across efforts, and leverage existing assets. They can do this because enlightened officials see the same problems that the public sees. That is, they do if the public makes a fuss and insists on the connection across all the issues we face.

IT'S ABOUT MOBILITY

Mobility is my real theme, and leadership is the real imperative. The goal is to move. Infrastructure is the platform that can slow us down and get us stuck, or speed us on our way. Mobility is opportunity. Mobility is freedom. Mobility is choices. Mobility and its patterns determine who gets educated, who can get to a job, who can take advantage of what cities have to offer, and who can make deals or relationships and with whom.

Mobility is independent of the particular mode of transportation. To get where we're going, we use whatever mode of transportation and cross whatever piece of infrastructure will get us to our destination. Fly between cities and then find a way to get into the city. Board buses and transfer between them. Get to work by car to the train, walk from the train station or hop on a bicycle from a bike-sharing station. Buy groceries, clothes, or office supplies that themselves might have traveled by ship to freight trains to trucks to cars.

We're multimodal people in a nation that tends to deal with only one mode at a time. It's hard to make progress that way, as paradigms shift and forces for change multiply. In this book, while analyzing infrastructure and transportation needs by mode of transportation, I argue continually for addressing the existing and potential connections across all modes. Indeed, better connections—say, plane to train at airports—or better attention to conflicts—say, bicycle lanes with separations that prevent a bumping up against cars—could make instant improvements in traffic congestion, even before technology is added that enables shared cars to be summoned on demand or automated cars to find their own parking. Of course, none of this is possible without leadership to craft strategies and mobilize support.

With connections in mind, I still have to start the journey somewhere, so chapter 2 examines rail transportation, the mode that built the continental United States into an economic power in the nineteenth century but has been split in recent decades between world-leading freight rail and languishing passenger rail. Chapter 3 looks at air transportation and its status both in the air and on the ground—at new navigations and communications technology that promises to create the quintuple wins, including coping with weather delays and air turbulence. Chapter 4 tours an exciting emerging ecosystem of roads + vehicles + users + networks that is "smart" and getting smarter, through the power of information and communications technology and the imagination of entrepreneurs with new business models. "Smart" roads, equipped with sensors, are dynamic partners with connected vehicles, including autonomous cars, and users are empowered with smartphone apps for ride sharing, car sharing, and parking. Chapter 5 makes the case for rethinking cities for the twenty-first century and asking cars to move over to make room for returning streets to the people, that is, to pedestrians, bicyclists, and bus riders, while downtown streets are beautified—a beauty-and-the-bus scenario. The closing of transportation disparities can make transit a ride out of poverty. Chapter 6 tackles the fundamental questions of leadership: politics and money, the will and the wallet. There are traditional and still-viable sources of finance, such as municipal bonds, and proposals for new models, such as regional infrastructure banks, but the most promising are public-private partnerships. Private investors can be attracted to public infrastructure projects when leaders with strategic visions get the politics out of the way.

Chapter 7 presents a call for a new, inclusive conversation that should be repeated throughout the nation, one that increases awareness, identifies priorities, stimulates innovation, and encourages action. A good way to start is to shake up our thinking, like twisting a kaleidoscope, to reframe tired policies and open new connections with the ideas and innovators building the future.

DO WE WANT twenty-first-century transportation and infrastructure to be people centered, technology enabled, environmentally friendly, opportunity focused, safe, and efficient? It's time to get moving.

2

ON THE RAILS

Are you hoping for futuristic, high-speed, closely connected trains enabling you to hop on and off for really fast trips, then connecting to planes or buses with a wave of your smartphone bar code? Sorry, the United States doesn't have that. How about just the high speed? Nope, we don't have that either. To find the best in trains, you must cross oceans. Here's what other nations have that America doesn't.

We don't have bullet trains. Japan's high-speed bullet train network, the Shinkansen, operates at speeds between 150 and 200 miles per hour. Launched in 1964, it covers nearly 1,500 miles. Shinkansen ridership was the world's highest (peaking at 353 million in 2007), until 2011, when ridership on the Chinese high-speed rail system surpassed it. The Shinkansen is totally separate from slower rails, uses a wider track than older lines, has no at-grade road crossings, and uses viaducts and tunnels to go over or through, rather than around, geographic obstacles. In 2012, the average delay from schedule was an amazingly minuscule thirty-six seconds. There have been no fatal collisions or derailments in its fifty years of operation. Advanced technology like automatic earthquake detectors and antiderailment devices can bring trains to a halt quickly when necessary. All-electric trains receive their power from overhead lines. The trains' axles are all powered, allowing for greater acceleration and traction compared with traditional loco-

motive propulsion, and automatic train control technology elimi-
nates the need for wayside signaling.[1]

We don't have maglev trains. "Magnetic levitation" trains use high-
powered magnets for both lift and thrust, rather than wheels on
tracks, ensuring low friction, the capacity for high acceleration, and
a smooth ride. Very expensive to build, only two maglev systems are
operational today. In China, the Shanghai Transrapid service, introduced in 2004, provides ultra-high-speed (268 miles per hour) passenger service between downtown Shanghai and Pudong Airport.
In Japan, the Linimo low-speed maglev was built for Expo 2005 and
functions like a light rail system. In July 2014, South Korea was set
to unveil its own maglev, between downtown Seoul and Incheon
Airport.[2]

We don't have high-speed rail over long distances. The French
TGV (Train à Grande Vitesse) introduced high-speed passenger
service of over 155 miles per hour to Europe in 1981 with a route
between Paris and Lyon. Since then, the system has expanded to
serve most of the country with about 1,300 miles of track; today
the trains routinely reach 200 miles per hour in regular service.
There are tilting mechanisms to manage turns and streamlined
locomotives at both ends of the train. High-speed rail has spread
to most of western Europe and the UK, as well as to Turkey, Russia,
and Iran. In Asia, high-speed rail networks have been introduced
more recently. Korea launched service between Seoul and Busan in
2004. Taiwan opened a high-speed line in 2007. In that year, China
launched its first high-speed service, and in 2012 it inaugurated the
1,428-mile Beijing–Guangzhou–Shenzhen–Hong Kong route, the
world's longest high-speed line. China now has the world's largest
high-speed rail network, totaling more than 6,900 miles.[3]

Here's what we do have in America. For passengers, it's not a happy
story.

We have one-size-fits-all tracks that are shared by freight, intercity passenger, and commuter rail operators. Virtually all the right-
of-way alignments in the United States were surveyed over a century

ago, bedeviling trains of all kinds with curves that require slower speeds. Amtrak's Acela Express "high-speed" service between Washington, D.C., and Boston is able to achieve its top speed of 150 miles per hour only for a very short stretch in Rhode Island and Massachusetts. Over its entire Northeast Corridor route, Acela must steer through eight different commuter operations over shared infrastructure vulnerable to delays and breakdowns, as well as tracks that are too close together for the heavy Acela trains to get up to speed safely.[4]

We have at-grade crossings in which roads and railroad tracks intersect, causing numerous delays, backing up traffic of all kinds of vehicles, and sometimes sending ripples far beyond the local area. Grade crossings are not only a congestion problem but also a safety hazard. In 2013, an Amtrak train hit a car or person at a grade crossing an average of 2.2 times a week; the figure is higher for other railroads.[5] As the rail hub of America, Chicago is notorious for dysfunctional crossings, which recently kept freight from moving all over the country; but other cities have similar tangles.

America's busiest passenger corridors are like working museums exhibiting artifacts from the early days of rail a century or more ago. There are overhead wires and other elements dating from the early 1900s, sometimes taking a decade or longer to repair or replace. Manual switching systems, often unidirectional, are only beginning to be automated with computer controls. Some aging components can fracture or break, meaning that a tiny piece can derail trains, cause accidents, and shut down the system.

While other nations have big, fast, and modern, we have small and quaint. Banking on nostalgia for the glory days of train service, Pullman Rail Journeys, a subsidiary of Iowa Pacific Holdings, offers twice-weekly trips between Chicago and New Orleans in refurbished Pullman cars, charging $2,850 for a luxurious master bedroom suite with a private bathroom and shower.[6] While such services can be incredibly fulfilling for those who can afford them, they are hardly a mass mobility solution.

The good news: we also have one of the world's most efficient freight rail systems, moving goods bound for our workplaces and households at low cost and in ever-more environmentally friendly ways, saving money for American consumers. Freight rail increasingly substitutes for trucks on American highways, reducing congestion for commuters and other road users. Managed as for-profit enterprises, freight railroads routinely reinvest 15–20 percent of their revenues back into their businesses. They have to. Among the most capital-intensive industries, America's freight railroads own, maintain, and repair their own tracks, and they pay taxes on the property and improvements. The *Economist* hails the "quiet success" of freight rail in America, citing the system's unparalleled cost effectiveness.[7]

That we manage to have a world-leading freight rail system despite numerous infrastructure problems and mediocre passenger rail can seem miraculous. How did we get to this place?

THE WEIGHT OF THE PAST: A 200-YEAR SAGA

In the original version of the popular board game Monopoly, a railroad was among the best assets a player could have. That was also true for much of American history, when owning a railroad guaranteed a monopoly—until trust-busting, bankruptcies, nationalizations, and then the ascendance of other forms of transportation in the mid-twentieth century pushed rail to the periphery, its tracks on the wrong side of history. At the close of World War II, the United States passenger rail system began to fall behind other developed countries. While Europe and Japan rebuilt, reinvested, and reinvented their war-torn passenger rail systems, the United States directed public investment toward highways and aviation. The result seventy years later is a distressed passenger system outranked in nearly every metric even by emerging-market countries, while freight chugs along on aging infrastructure.

Fluctuating fortunes and political battles figured in the railroad story even earlier in history, almost to its origins nearly two hundred years ago. Railroads have always been characterized by hybrid organizational models (not quite private, not quite public) and complex cross-sector stakeholders. But some people still ask about the Transcontinental Railroad, imagining a nineteenth-century golden age of national infrastructure investment in the public interest. Even then, when rail was the advanced transportation mode of choice, development was possible only because of a temporary alignment of multiple interests, and incentives that sometimes encouraged corruption. As the joke goes, it was a golden age only for those who got the gold.

Between the Louisiana Purchase in 1803 and the Mexican Cession in 1848, the western border of the United States moved from the Mississippi River to the Pacific Ocean, a distance of over 1,700 miles, invoking a "manifest destiny" for continental hegemony. Defense of the western territories required settlement, which would also create economic opportunities. Railroads powered by steam engines could be the key to defending and developing the West. The idea was an early version of what twenty-first-century policy makers call "transit-oriented development"—that new access to rail transportation could transform previously worthless land into valuable real estate by allowing settlers to move west from eastern population centers, and agricultural products, precious metals, and other goods to move east. We see even today the economic benefits of train stations and subway stops; I still regret not having bought investment property in a slightly seedy area about to get a subway extension.

Despite defense and development benefits, several decades of discussion made clear to advocates for a Transcontinental Railroad that the project would not become a reality without some version of what would today be called a "public-private partnership." With an estimated capital expenditure of $100 million (roughly $2.8 billion in today's dollars) for only half of the project, unknown operating costs relating to safety and security in hostile territory, and few western population centers, the business case for a for-profit endeavor was poor. A fully public-run approach was likewise out of the question. Federal government construction and operation of a railroad was widely viewed as unconstitutional. And even if the new state and territorial governments in the West could coordinate among themselves, they lacked the management competence and technical expertise to deliver the most daring public works project the country had ever contemplated. Reluctantly, railroad entrepreneurs went to Congress for help. They found a legislative body broadly in favor of the project, but deeply (and fatally) divided on almost everything else. (Sound familiar, today's Congress watchers?) Discussions over the route of the railroad, and thus the economic development path, brought in the contentious pre–Civil War debate over slavery.

In 1860, Abraham Lincoln was elected president on a Republican Party platform that included the railroad, and in 1861 Congress passed the Pacific Railroad Acts, which facilitated granting land development rights. The Transcontinental Railroad was built after the Civil War, between 1866 and 1869, starting from both the East and the West Coasts. Labor shortages during the California Gold Rush and the aftermath of a devastating war meant that six thousand workers had to be brought from Guangzhou, China, to lay track. Management fiascoes abounded. To claim the largest federal subsidy, the two railroads headed toward each other with minimal coordination and actually passed each other; identifying a meeting point for the two sets of tracks required the mediation of both President Andrew Johnson and President-elect Ulysses Grant.

By 1880, about 93,000 miles of track had been deployed, compared with only 1,000 miles in 1835. The result was a dramatic increase in the density of coverage in the East and an expansion of reach in the West.[8] Despite attempts at government oversight and accountability measures, there occurred political wheeling and dealing, earmarks, inflated billing, labor exploitation, seizure of Native American land, and business corruption. Most prominently, the dominance of long-haul railroad transportation meant that railroads' inherent natural monopolies over the routes they traveled grew into a wave of monopoly power over suppliers and customers alike. Gradually, railroads also began to collude to ensure that markets would not be saturated with "too much" competition. Magnates like John D. Rockefeller, Andrew Carnegie, and William Vanderbilt brought railroads into "trusts" that also owned steel and petroleum companies. Railroad companies went to ruthless—even abusive—lengths to control labor sources, leading to a series of violent strikes in the late 1870s that threatened to disrupt the agricultural production and distribution system. Infuriated farmers organized the populist Grange political movement.[9] Public mistrust of business spiked. Politicians ran for office on trust-busting platforms.

To rein in the railroads, Congress passed the Interstate Commerce Act in 1887, creating America's first business regulatory agency, the Interstate Commerce Commission. Railroads then entered a period of investment in the early twentieth century. Pennsylvania Station in New York City and the tunnels, yards, ridges, and infrastructure that support its operations were completed in 1910. Electrification of the Northeast Corridor occurred during that time and was largely complete by 1930. "It's a good thing this was done," observes Marc Hoecker, director of strategic planning for a U.S. Class I freight operator, "because that's what we've been relying on for service in the Northeast Corridor ever since."

Following World War II, the railroad industry's decline accelerated. Postwar societal and technological changes made air and automobile or truck transportation feasible alternatives to rail. Innovation spurred by the Department of Defense—with defense long the rationale for national investment—was good for air transportation. Although the first commercial airlines in the United States began appearing as early as the 1920s and 1930s, the slow speed and high costs of their services made them poor competitors. But government-developed and government-funded technological advances during World War II changed that. In the decade following the war, Boeing, Douglas, and Lockheed introduced a new generation of commercial aircraft, largely based on design and features developed for the military. Jet engines came into use in commercial aviation, which dramatically improved the speed and cost-effectiveness of long-distance air travel for moving people quickly and cheaply.

Defense logic also helped roads prevail over rail. After World War II, a generation of military veterans returned from war eager to start families, and the resulting baby boom drove a major migration from cities to outer suburbs, where low population densities made economically viable rail service difficult. Automobile transportation, in contrast, was ideally suited for low-density settings. Government-subsidized road construction under the National Interstate and Defense Highways Act of 1956 created the Interstate Highway System. President Eisenhower championed this as a defense necessity for evacuating cities and moving troops. In 1919, when he traveled cross-country as a young military officer, the trip took two months; with the new highway system, it would take five days. Auto companies had long been aggressive in promoting roads used by cars and trucks over other forms of transportation. Advances in auto speed and safety, plus increased personal wealth thanks to the postwar economic boom, helped cars and trucks steal share from railroads.

Regulation the railroads found burdensome, including pricing constraints, perpetuated decades of deferred maintenance, under-investment, low-quality services, and poor financial performance. Passenger services fared far worse than freight. Railroads were nationalized twice, in 1917 and 1941, for war mobilization efforts, then privatized and shoved to the background in national policy. A low point came in the 1970s, as multiple railroads filed for bankruptcy, and Congress nationalized passenger operations with the formation of Amtrak. In 1980, nearly a century after the Interstate Commerce Act, partial deregulation under the Staggers Act trimmed the powers of the ICC and allowed railroad companies to make their own contracts with customers and to abandon money-losing lines. Ironically, that lowered prices on the freight side; adjusted for inflation, freight rail transport prices decreased by half from the 1980s through 2002.[10]

"Rail has been under attack for over fifty years, since the start of massive subsidies of highway and air," an industry insider declares. As demand for passenger services collapsed, railroads reduced services to the bare minimum, decreasing customer satisfaction and further depressing demand.

Thus, for over 150 years, the federal government stepped in numerous times to stimulate, nationalize, or reorganize the railroad industry. In the mid-nineteenth century, President Lincoln's administration created the Union Pacific to handle half of the Transcontinental Railroad and appointed two of its board members. In the first part of the twentieth century, railroads were periodically nationalized or placed under heavy regulatory control to support wartime mobilization for World War I and World War II, and twenty-five years later to keep the industry from collapsing. Amtrak (passenger rail) was formed as the main national passenger rail company under President Nixon in 1970 after the Penn Central bankruptcy, then the nation's largest business failure, tainted by an insider-trading scandal. To manage the chaos of the remaining restructurings, Conrail was created a few years later, in 1974, as a major government-owned freight rail system.

Amtrak was established as a hybrid of a for-profit and government entity: a company that sold stock and was expected to make a profit (which it didn't), while also a government entity dependent on Congress for budget appropriations and subject to government work rules. In 1997, President Clinton signed a reauthorization of Amtrak that provided a then record $2.3 billion in payments for capital improvements to the rail system, on the condition that operating subsidies be phased out. A new Amtrak board targeted high-speed rail as the centerpiece of its strategy. Neither of the passenger rail visions have been realized, let alone anything resembling the bullet and maglev trains of other nations.

Not much has changed in the twenty-first century, as Amtrak limps along with revolving-door management, undercapitalization, and a perpetual mission conflict: to provide a vital public service and make money at the same time. There have been six CEOs since 2002, each bringing reorganizations and new management teams.

For solutions to our rail struggles, we should look not abroad but to the other half of the history: the quietly strong American freight rail.

GETTING THE GOODS: THE GREENING
OF FREIGHT RAILROADS

Conrail served as the Northeast region's primary freight railroad company for more than twenty years. In 1999, it was sold to the private-sector railroads CSX and Norfolk Southern. Those companies further invested in freight rail, helping the industry get greener in three senses: the green of money from profitable growth, the green of improving environmental performance, and the potential for greener and less-congested highways by replacing trucks as the main carrier of goods. Who knew that those noisy freight trains looking brown and rusty could be so green?

Michael Ward knows. Ward loves railroads. His loves his own railroad company, CSX, which traces its origins to 1827 when the Baltimore & Ohio Railroad was formed as the nation's first common carrier. He traces his own origins at CSX back thirty-seven years, when he took an analyst job as a newly minted Harvard Business School M.B.A., rising to become chairman, president, and CEO in 2003. And he loves the whole American freight rail industry.

"Railroaders are like farmers," Ward declares. "You heard about the farmer that won the lottery? They said to him, 'Oh my gosh, you won the lottery; what are you going to do with all that money?' He said, 'I'm a farmer and I love farming, and I'm going to farm until every penny of it is gone.' And I say railroaders are like that. When we make more money, we're going to invest more back into the infrastructure, so we can strengthen the railroad and grow the business."

Ward may sound like a press release, but that's exactly how he talks, and why he's a major industry spokesman. He lavishes praise on industry performance: "While we've improved the profitability of the industry, we've also cut rates in half of what they were in 1980 for our customers, on an inflation-adjusted basis. We're providing a more economical product to them, and it's safer and more reliable. Over the years, as an industry, our train accident rate is down 80 percent; our personal injury rate is down 85 percent; and we're doing this with about one-third of the workforce we had in 1980." He calls the industry "the envy of the world."

CARRYING THE LOAD: THE FREIGHT RAIL INDUSTRY

In 2011, the American freight rail system transported 1,680 billion ton-miles of freight, accounting for 25.4 percent of global volume. While the market for freight rail services in northern Europe is projected to shrink from $32.1 billion to $30.8 billion by 2016, the market in North America is expected to grow from $29.5 billion to $33.5 billion. With a railroad network stretching 139,679 miles, the reach of America's infrastructure is second only to the European Union's (by less than 2,485 miles) and is nearly three times larger than that of Russia, the third-ranked country. Freight rail operators in the United States handle 40 percent of the country's freight movement by distance.[11]

Few other developed countries have freight rail systems offering the same level of choice and competition as America's. For example, the French government recently announced plans to reconsolidate ownership, access, and maintenance of rail infrastructure from three separate companies back into one large state-owned enterprise, increasing difficulties for independent operators. Incumbents dominate freight rail transportation in continental Europe; principal operators control over 70 percent of the market in Germany and Italy, 80 percent in France, and over 90 percent in Spain. Absent a unified European market, a freight train traveling the 980-mile journey from Slovenia to Istanbul will change locomotives eight times and cross five countries—each with its own rules about track access fees, operator licensing, voltage, gauge, and safety and signaling systems.[12]

U.S. freight railroads come in three varieties: large, medium, and small. Seven long-haul Class I railroads (the feds do the classifying) with nationwide reach form the backbone of the freight rail system. They operate 162,000 miles of primary and secondary track, owning most of it, control about 35 percent of the freight cars in service, and account for 90 percent of the industry's revenues. CSX and Norfolk Southern rank third and fourth in revenues, below Union Pacific and BNSF; in 2010, legendary investor Warren Buffett showed his faith in the future of freight rail when his Berkshire Hathaway bought a major stake in BNSF. In 2012, the seven big freight railroads collectively moved 1.76 billion tons of freight, primarily coal (41.0 percent by volume), chemicals (9.9 percent), farm products (7.9 percent) and minerals (7.4 percent), and employed 158,623 people.[13] Most freight lines run east to west, connecting the coasts, but Kansas City Southern, the smallest Class I railroad, runs north to south across the Midwest. It is known as the NAFTA railroad (after the North American Free Trade Agreement) because its lines penetrate deeply into Mexico and, with partner Canadian National, into Canada to connect the large, promising North American market.

Twenty-one Class II regional railroads link their much bigger cousins and short-line railroads. A supporting cast of 539 Class III railways connects local manufacturers to the larger rail system or trucks to trains, to and from ports or intermodal truck terminals, a rapidly growing business. Some of these are owned by holding companies and networks, including a consortium owned by Union Pacific, BNSF, CSX, and Norfolk Southern; mergers are also consolidating the industry. Watch more investors jump aboard.

Class I railroads have been able to maintain relatively high levels of profitability (and recent margins of 17.1 percent), even through the economic downturn. Revenues nearly doubled between 1998 and 2008, to $61.2 billion, when the global financial crisis hit and reduced demand. But prices increased, and revenues rebounded to over $70 billion in 2012. Class I railroads have aggressively reinvested profits in infrastructure, increasing annual capital expenditures from $6.4 billion in 2005 to $11.6 billion in 2011.[14]

The pattern of financial profitability and reinvestment of profits in the improvement of infrastructure helps freight railroads contribute to the quintuple wins. Rail is widely considered the most economically efficient surface transportation mode for moving goods, and the most environmentally friendly. According to the American Association of State Highway and Transportation Officials, the freight moved on trains would cost shippers an additional $70 billion to move by standard trucks.

WHY WE CAN'T KEEP ON TRUCKING

A hidden secret of how you get what you need: sometimes shipments of goods that appear to be carried by trucks are actually moved largely by rail, as truckers take them to railroads for transfer between major cities and then pick them up at the other end for delivery; the trucking companies handle the contracts with the shippers and receivers, then subcontract to freight railroads. With the advent of containerized shipping, the same container can go on a ship, a truck platform, or a rail flatcar. This makes it efficient to transfer cargo from one mode to another. But it doesn't explain why trucks would turn to rail.

Driver shortages are one reason. For some companies, turnover of drivers reaches 100 percent per year; at one company, the only positions with a higher turnover rate is the director of driver recruitment, an industry insider says. That old-fangled driver lifestyle of weeks away from home on the road just isn't appealing to as many people these days. The frustration of highway congestion can mean frequent delays and higher costs. Rail and truck are no longer competitors; they are increasingly allies. At the summit I convened, I watched Michael Ward of CSX nod at former Kansas governor and American Trucking Associations president Bill Graves a few seats away, as he said with a smile, "We're actually partnering with the trucking industry now. We can provide some good solutions both economically and environmentally."

The truck issue has taken on greater urgency with public awareness of the dangers of driving while drowsy. On June 7, 2014, come-dian Tracy Morgan was seriously injured and a companion killed on the New Jersey Turnpike when a Walmart truck, whose driver allegedly hadn't slept in more than twenty-four hours, crashed into their van. Although regulations have officially limited the hours truckers can drive, in practice, numerous accidents have been attributed to drivers' falling asleep at the wheel. "In a business that lives by the clock, miles mean money. Commercial truck operators have resisted, arguing, in effect, that Washington cannot regulate sleep," suggested a *New York Times* report.[15] But regulation of driver hours is well under way, with electronic tracking systems tightening the screws.

Taking trucks off the road can have a dramatic impact on safety and quality of life. Norfolk Southern chairman Wick Moorman points to the Crescent Corridor as a positive example. The Crescent Corridor is a $2.5 billion-plus rail infrastructure project that spans eleven states with fast, direct routes from Louisiana to New Jersey and convenient connections to Mexico and Los Angeles. Here's the lifesaving part: "Along the Interstate 81 corridor, truckload traffic was two and a half times what was expected on the highway when it was designed. Churches were adopting miles of highway to pray for the motorists," Moorman says. "We started to talk with states and with federal officials. We generated a plan to take 800,000 to a million trucks off the highway. Ultimately, the project saved half a billion dollars in trucking costs, half a billion dollars in congestion costs, took a million trucks off the highway, and reduced heavy truck crashes, all with nearly two million tons less carbon."

With thirty new lanes now open that connect to the 2,500-mile Crescent Corridor, Norfolk Southern's high-capacity intermodal routes are on track to make green shipping a reality, as the company proclaims, which means fuel efficient, and dependable. By the time the project is completed, there will be 122,000 new jobs across the rail network, new terminal facilities in Birmingham, Alabama, Memphis, Tennessee, Charlotte, North Carolina, and Greencastle,

Pennsylvania, and over 1.3 million trucks off the roads. This kind of project is en route to achieving the quintuple wins.

Other regions haven't always caught up. The Alameda Corridor project in California involves building a rail/highway separation to move more freight by rail from the ports of Los Angeles and Long Beach to inland destinations. The idea originated with the railroads and port officials impatient with the slow pace of Caltrans, the state DOT. Rail traffic is increasing, but too many trucks remain, causing congestion and pollution, according to a consultant's report.

INSIDE THE FREIGHT PROPOSITION:
HOW A FREIGHT RAILROAD DELIVERS

Rail is efficient because it uses less energy than other land-based transportation modes, and that also reduces pollution and lowers greenhouse gas emissions. Freight railroads claim to be four times more environmentally friendly than other modes of transportation. Moving freight by rail is three times more fuel efficient than moving freight on the highway. Trains can move a ton of freight nearly 450 miles on a single gallon of fuel. (Fuel efficiency for trains is measured in terms of ton-miles, because the length and the weight of trains vary greatly.)

Take a look at Michael Ward's company, CSX. CSX delivers freight that reaches nearly two-thirds of American consumers through a transportation network of 21,000 miles across twenty-three U.S. states and two Canadian provinces. Its purpose and vision, the company proclaims, is to capitalize on the efficiency of rail transportation to serve America, and to be the safest, most progressive North American railroad, which includes a community service and employee empowerment ethos. It declares that every $1 billion CSX spends to improve the efficiency and safety of its network helps to create 17,500 new jobs. The railroad funds its own repairs and owns its tracks; 16–17 percent of its revenues are spent every year on capital improvements, about $900 million of which is for the tracks. In addition, about $700 million in annual operating expenses goes into the track.

In addition, CSX has invested $1.5 billion over the past decade to improve its locomotive fuel efficiency and reduce the corresponding emissions, working with manufacturers. "We're partnering with GE to test if liquefied natural gas will make sense for the rail industry. The challenge will be putting the infrastructure into fuel and making the retrofit investments. The toughest decision is not the technology but the bet on the price movement of two fuels, which is a tough decision to make when a huge capital investment is required to retrofit the locomotives," Michael Ward says.

To reduce fuel consumption and carbon emissions during idling, CSX has invested in two separate pieces of idle-reducing technology, Auxiliary Power Units and Automated Engine Start Stop. Auxiliary power to a locomotive allows the larger diesel engine to be shut down, and the automated system shuts the locomotive down when not in use and automatically starts it when needed. Any of us who drive automobiles will immediately grasp why this is important— and we might want such a device for our cars. The company also uses a system similar to a "black box" on an airplane, in order to monitor and record actual train operations; that gives the engineers feedback on how to improve fuel efficiency and operations in general. Some of those old-fashioned freight trains have more advanced technology than the newest automobiles.

As a result, CSX is recognized as a top ten company in the S&P 500 on both the Carbon Disclosure Leadership Index and the Carbon Performance Leadership Index. CSX is the only railroad listed on either disclosure index; its score of 95 was the fourth-highest on the fifty-three-company S&P 500 carbon disclosure list and second-highest for S&P industrial companies. The score also placed the company fourth among all global industrial companies. A 2013 U.S. Environmental Protection Agency Climate Leadership Award in the Goal Achievement category recognized CSX's 8 percent voluntary reduction of greenhouse gas emissions intensity since 2007.[16]

In 2013, CSX achieved industry-leading safety performance, as well as record-high customer satisfaction, revenues, and earnings per share from continuing operations, all while continuing to invest in infrastructure for the future. All this helped make CSX number 2 in its category (trucking, transportation, and logistics) on *Fortune* magazine's World's Most Admired Companies list in 2014. The list measures corporate reputation on the basis of attributes ranging from social responsibility and innovation to long-term investment value and product quality. As if that weren't enough, for two years running, in 2013 and 2014, it was number 1 on DiversityInc's list of top companies for veterans. Even its police department receives honors, alone among railroad companies.[17]

Community service is important to companies that depend on community support for work on the tracks running through them. But Ward's support for City Year, an education-focused urban youth corps on whose national board he and I both serve, goes far beyond PR benefits to CSX as a national sponsor. He and his wife, Kim, donated millions of dollars of their own to bring City Year to Jacksonville, where CSX is headquartered. It is part of CSX's desire to improve the prospects for disadvantaged communities that lie along its tracks—you know, the places once known as "the other side of the tracks."

My point in reciting this litany of awards is to show that even the grubbiest, oldest of industries can lead the way in modernizing, can run values-based companies, and can apply new technology. I highlight CSX as one example, but the pattern is not unique to CSX. Class I railroads and the industry have adopted the same or similar initiatives and collectively set records in operational and financial performance. There's a lot we can learn from the freight rail mode, and it's worth keeping this in mind later, when we turn to passenger rail.

ALTHOUGH HEALTHY AND viable, freight rail is not flawless. One high-profile controversy is over the best way to move oil, whether by pipeline or in railcars. The Keystone Pipeline, which would move synthetic crude from the oil sands of Alberta, Canada, and crude oil from the northern United States to refineries primarily on the Texas gulf coast, is hotly debated in Washington, as Phases 3 and 4 are yet to be funded; environmentalists are particularly opposed to the pipeline. However, as more oil is carried by rail, fatal accidents have increased. Congress and regulators are looking closely; the industry collaborated on new tank car standards and agreed to decrease the speed of freight trains crossing cities. High speed might be the dream for passenger rail, but freight chugs along more slowly.

A high-profile setback occurred just across the border. On July 6, 2013, an unattended 74-car train carrying crude oil ran away and derailed in Lac-Mégantic, Quebec, leading to at least forty-seven deaths and a serious fire destroying more than thirty buildings, roughly half the downtown area. This was one of the deadliest rail disasters in Canadian history. In August 2013, Canada's transportation agency suspended the license of the U.S.-based company, the Montreal, Maine & Atlantic Railway, and its Canadian subsidiary operating the train. It said that the disaster had raised questions about the growing use of rail transport for oil, including important ones regarding the adequacy of third-party liability insurance coverage to deal with catastrophic events, especially for smaller railways. A year later, the railway, now under new ownership and named the Central Maine & Québec Railway, has once more commenced operations in Canada, although it will not be transporting any crude oil through Lac-Mégantic until at least 2016.

But outdated infrastructure, equipment, and labor practices were also pointed to as possible culprits—the same issues that plague passenger rail. Added to that is an irony of success: rising demand for freight rail services, with capacity falling behind.

On September 10, 2014, the U.S. Senate Committee on Commerce, Science, and Transportation held hearings about improving the performance of America's rail system. Committee chairman Senator John D. Rockefeller IV (D-WV), whose inheritance stems in part from railroads, opened with this statement:

> The freight railroads continue to set new financial records quarterly, and these companies continue to raise their dividends and

buy back record amounts of stock. But not everyone is doing so well.... For several months now, the agricultural, coal, chemical, and automotive industries, among others, have been experiencing serious rail service delays, sometimes on the order of months. And it's not just industry—passengers are also feeling the effects. Amtrak's long distance trains around the country are being severely delayed. Whether it's been extreme winter weather, a surge in Bakken crude oil production, a recovering economy, or a combination of factors, we must do more to move our grain to market, coal to power plants, automobiles to consumers, and passengers to their destinations. For many shippers, this is their livelihood, and it is too important to not do anything.

What can be done to keep freight rail delivering the goods?

THE ENGINES THAT CAN: FREIGHT RAILROADS AS REGIONAL CHANGE CATALYSTS

The Little Engine That Could, a classic children's book, tells a very American story of how a small person with grit, determination, and compassion can stretch to get something done for communities that the big players ignore. A locomotive carrying a trainload of toys bound for children on the other side of a steep mountain fails and pleads with passing trains to complete the journey. After big, powerful trains refuse to help, a very little engine agrees to try, although it is concerned about whether it can make it. The little engine chants, "I think I can, I think I can ...," as it chugs up the mountain, until it says, "I know I can," and the toys arrive to delight the children. It's a freight rail parable for today.

That's the surprise. Leaders from America's freight rail companies are partnering with government entities to apply scarce resources in innovative ways to remove bottlenecks and modernize infrastructure, so that goods and people can get where they need to go faster and more efficiently today, while creating new capacity for a future already in sight. For one thing, the expansion of the Panama Canal, anticipated to be completed in 2015, once finished will bring bigger ships to and from East Coast ports, carrying massive loads of cargo to be transported from and to inland destinations.

On the passenger side, a growth in demand for light rail and higher-speed trains fits the values of an environmentally conscious, non-car-owning younger generation. Passenger rail, whether long-distance or light rail for commuters, must be part of the mix in strategies to reduce highway congestion and unclog city streets. And if tracks are neglected or abandoned, it will be hard to get them back as the land is used for other purposes.

A quarter of a trillion dollars is needed to restore America's passenger rail infrastructure, according to OECD estimates. That's a daunting figure, big enough to tempt even the strongest advocates of rail to give up. Abandonment of aging, crumbling areas is also an American way, as evidenced by blighted areas of bankrupt cities like Detroit. But between the impossibility of big spending and the disaster of continuing neglect could lie a third way, through the creativity of little engines willing to pitch in, and reflecting U.S. can-do spirit.

Michael Ward, Wick Moorman, and their companies, CSX and Norfolk Southern, are hardly little, but they are a force for change different from any that has appeared in railroad history in the last century and a half, when the federal government made things happen with bold strokes that turned the industry on its head. They also bear no resemblance to the financial moguls who brought the now-busted trusts of the late nineteenth century; Michael Ward and CSX might be philanthropic, but they're no J. P. Morgan. The CSX and Norfolk Southern teams work from the ground up where they see the needs. Using their imaginations to innovate, they also reach into their pockets to contribute and understand that they must improve the communities around them as well. They act not alone but as members of coalitions, cooperating even as they compete, with local and regional governments as players and even initiators of collaborative projects. Coalitions are not like the busted trusts; they are more inclusive, more transparent, more temporary, more participatory, less power seeking. This way of working across sectors prods rigid bureaucracies to get moving. Local and regional action can remove pain points and bottlenecks through collaborative leadership, simple innovations, and new financing mechanisms.

Consider two compelling examples of how freight railroads are helping solve regional problems that have great national significance—not only for shippers but for all of us waiting for goods and good food, or to board a passenger train.

THE SLOWEST SIX MILES IN AMERICA

What happens in Chicago doesn't stay in Chicago. When some kinds of local infrastructure problems are unaddressed, the consequences ripple across the nation. It's not just when the trains can run; it's whether anyone can get anywhere and get what's needed. Dysfunctional infrastructure left over from past decades is a major detriment to quality of life.

Over nearly 180 years, Chicago evolved to become the most important railroad center in North America. The city serves as a critical exchange point for the nation's freight rail traffic, with six of the country's seven largest carriers accessing the region. A quarter of all freight trains in the United States will travel through Chicago at some point in their journey. The city is also the primary hub for intercity passenger rail transportation. The region's commuter rail services are the second largest in the country by volume.

Unfortunately, Chicago's position as a hub for rail services also makes it a bottleneck. Three of the ten most congested rail spots in America are in Chicago. (If you're curious, two others are in Houston, and one is in Fort Lee, New Jersey, site of the 2013 bridge scandal.) For Chicago, a century's worth of deferred investments meant that by the year 2000, a train that took less than forty-eight hours to travel 2,200 miles from Los Angeles could then take an average of thirty hours to cross Chicago, at a speed that is equivalent to "a quarter the pace of many electric wheelchairs," as a journalist said. Effects from any bottleneck in Chicago cascade throughout the country's freight rail system. A particularly severe blizzard in Chicago in 1999 created a national freight backup that took months to clear.[18]

Freight rail congestion also creates problems for passenger rail travelers that ripple out to communities in sometimes dangerous ways. In Illinois, Amtrak and intercity passenger trains face frequent delays because of the density of trains on the tracks, despite being given priority use over freight trains. Although the region's commuter rail operators own their own infrastructure, and passengers take precedence over freight, their tracks often converge at grade level, requiring commuters to wait while miles-long freight trains slowly cross their paths. In Chicago's suburbs, long freight trains could block the flow of cars—and, ominously, emergency vehicles—for hours. One local mayor encouraged residents to cross the tracks the first chance they got, because they might not get another chance for a very long time.

Much of the area's congestion is due to counterproductive interactions between freight rail and other components of the transportation system. The seemingly haphazard crossing of tracks at various angles was a product of all that had happened to rail in the twentieth century. Investment in rail was not a public priority, and there was no real strategy, so private operators laid track wherever they wished. The industry was ignored, nationalized, consolidated, and then picked apart again, reducing motivation for maintenance or cooperation. Eventually, instead of many short trains from many independent railroads, a few big surviving operators began running longer trains, which held up everything.

In the aftermath of that disastrous winter of 1999 and an effort to reposition freight rail infrastructure as a source of competitive advantage for the region, rather than a pain point, discussions began among a range of local leaders. In June 2003, under the powerful then Mayor Richard Daley, the Chicago Region Environmental and Transportation Efficiency Program (CREATE) was launched. (Transportation people love acronyms.) A principal goal of CRE-

ATE was to eliminate the dysfunctional grade crossings where railroad tracks intersected roads and each other, and freight trains interacted with cars, trucks, or passenger rail at the same physical level. The CREATE team had representatives from federal, state, and local governments, including suburban communities. The $3.2 billion umbrella program, with seventy projects that could be completed one at a time, included twenty-five new roadway overpasses and underpasses as well as six new rail overpasses and underpasses, all aimed at segregating freight rail from the rest of the system. It also made a nod to modernization: it overhauled the region's hand-operated switching system by installing a computerized traffic control system that allowed trains to travel faster and more safely.

In an exceptional gesture, six fiercely competitive freight rail carriers, including CSX and Norfolk Southern, plus Amtrak on the passenger side, put aside their differences to share technical information, coordinate capital project planning, and work together on streamlining operations in the Chicago region. One result: the Common Operational Picture tool, which integrates information from the dispatch systems of all rail users. By providing a real-time view of all 1,300 freight and passenger trains on the region's rail infrastructure in a given day, the tool (known, naturally, as COP) helps operators to proactively identify congestion and reroute trains as needed, to move people and goods through the Chicago area quickly and efficiently.

This is long-termism, for sure, transcending political administrations and seeking continuing funds through economic ups and downs. Mayor Rahm Emanuel inherited CREATE when he was elected in 2010. But as Deputy Mayor Stephen Koch said on my visit in May 2014, elected officials find it difficult to get in front of voters with long-term propositions that might cost taxpayers money. Vocal support from longtime private-sector civic leaders and associations, plus the cooperation of the railroads, can provide cover for those who run for election every few years. Ann Drake, CEO of DSC Logistics and vice chairman of Chicago's Metropolitan Planning Council, puts "funding for CREATE" at the top of her wish list for improving transportation in America. Moreover, without CREATE, Chicago would lose valuable manufacturing jobs, because manufacturers want access to efficient freight rail. This was proven to local officials when Ford threatened to close an auto assembly plant with 1,600 high-wage jobs. As of this writing, thanks to the promise of CRE-ATE, Ford has 4,000 jobs, with suppliers located nearby.

I visited the area near the Ford plant, at 130th Street and Torrance and Brainard Avenues in Hegewisch, in Chicago's Tenth Ward, one of seventy-seven community areas, on the far South Side near the Indiana border. Officially known as GS15A in CREATE-speak, it is dubbed a "911 critical crossing," and I could see why. It was a mess.

We drove there from downtown Chicago in a Chicago Department of Transportation van with Joe Alonzo, a Chicago DOT planner working on CREATE, Jeffrey Sriver, a coordinating planner, and Soliman Khudeira, project director. We passed other CREATE sites on the way, finished or under way, but this site, still under construction, was the most complex. I imagined that previously it was like a nightmarish novel, in which everything converged from all directions to a point with no exit.

The area was a daily traffic jam of wild proportions: two hundred hours a day of vehicle delays, the Chicago DOT staffers said, for the 32,000 cars and trucks that passed through. To add insult to injury, this already overloaded area was also a designated truck route. Two dozen trains a day were responsible for the congestion; sometimes a long one could take twenty minutes to pass by. The tracks were everywhere, at odd angles. We stood on the edge of the new road-bed and could see it all, including the bright blue bridge, a so-called instant bridge that had been assembled elsewhere and rolled into place. On one side of the construction gravel, I could see the empty marsh; on the other side, an existing road and, beyond it, the Ford plant, with sky bridges between its two sections. Ford still had to get finished cars over the tracks to the parking area by the New Car Shipping Center, from where the cars could move by train or truck to national and global destinations. Employees still had to use the old roads to get to work, which must have been hugely frustrating.

The GS15A project is the piece of CREATE that is aimed at a Norfolk Southern grade separation in the area of the Ford plant. The work involves an unusually large number of components: lowering two roads so that bridges could move trains overhead, adding three railroad bridges (two for freight trains, one for commuter rail), building a new roadway bridge and two pedestrian bridges, relocating train tracks, and creating bicycle and pedestrian paths along 130th Street, where a park will emerge out of marshland. There will also be new drainage ditches, traffic signals, and the like.

The rubble and the remoteness reminded me, I'm sorry to say, of the many poor Third World countries I've visited that were just developing. It's hard to get excited about something that shouldn't have been neglected and allowed to become such a mess in the first place. But there were many things to applaud about the leadership courage to get this going, and the planning and engineering talent to execute. Someday soon, this will be a vibrant industrial and residential hub with speedy travel connections. No longer will blizzards delay the movement of stuff by freight through Chicago and on its way to the rest of America.

Delivering GS15A and the rest of CREATE's seventy projects and ensuring that improvements are sustained requires a new way of working, particularly when it comes to managing the rigid environmental impact requirements of the National Environmental Policy Act of 1969. Using a typical environmental review methodology, each component project of CREATE would be evaluated individually and sequentially, a process that could take years. Recognizing that this would be a major barrier to private-sector participation, federal, state and local leaders created a new process called the Systematic, Project Expediting, Environmental Decision-making (SPEED) strategy. (Acronym cleverness strikes again.) The SPEED strategy provided "an expeditious method of moving low risk component projects forward," while assessing environmental impact in a "proportional, graduated" manner. It's nice to know that the bureaucracy can invent bureaucratic ways to move faster.

By mid-2014, CREATE was well on its way to reducing congestion in America's biggest rail hub. Twenty of the seventy CREATE projects had been completed, and more than $1.2 billion had been spent or committed to the program. Several engineering milestones had been achieved in the process, including the rolling of a 4.3-million-pound truss bridge—the largest ever preassembled off-site—into place. Some of the region's most notorious grade crossings of road and rail had been eliminated, and delays had been reduced for thousands of vehicles per day.[19]

Project-by-project success doesn't mean that the whole job will get done. It is unclear where an additional $1.7 billion would come from to continue CREATE projects after 2015. Disagreements among local stakeholders over whether it is the responsibility of the federal government or of local agencies to find additional financial contributions increases uncertainty for CREATE's private-sector partners. Advocates face the challenge of finding an equitable and sustainable funding source for a project that is clearly having an impact on efficiency and competitiveness.

Fifteen years after it was first envisioned, CREATE offers a promising example of progress when parties come to the table to work together and use resources already available in a strategic, focused, and integrated way, providing demonstrable results that everyone cares about.

NATIONAL GATEWAY GRIDLOCK: LEAPING FROM THE NINETEENTH TO THE TWENTY-FIRST CENTURY

Not every improvement needs to be complex or carry a billion-dollar price tag. Sometimes simple innovations can have a dramatic impact. The National Gateway initiative led by CSX serves as an example of leveraging existing assets by focusing attention on repair and adding simple, relatively low-cost innovations.

Freight rail has long faced its own physical form of "Washington gridlock" in the city's Virginia Avenue Tunnel. Built by the Baltimore and Potomac Railroad seven years after the end of the Civil

War, the 4,000-foot tunnel remains a critical pathway for freight trains bringing manufactured goods from the Midwest to the Port of Baltimore, part of the National Gateway Corridor. Despite its importance, low overhead clearance—another legacy of the nineteenth century—makes the tunnel a choke point for freight services.

Using freight rail in this area is a no-brainer. Except for one big problem: not enough capacity to handle the traffic. Stretches of single-track infrastructure could become major bottlenecks, and the building of additional tracks would be prohibitively expensive. CSX and the freight rail industry saw a solution in a late-1970s technology called "double-stacking," which allowed the same rails to deliver twice the throughput. Instead of placing a single container on a railcar, a new generation of freight terminals could stack *two* containers on one specialized railcar. In addition to using rail infrastructure more efficiently, double-stacking offers major economic and environmental benefits. A double-stacked freight train is twice as fuel efficient as a regular train and more than five times more efficient than a truck.[20]

Accommodating double-stacked trains takes investment, but not as much as the alternatives, and it is in the affordable range. Beyond new terminals and cars, the increased height of double-stacked containers compared with traditional freight railcars requires raising the ceilings or lowering the floors of existing tunnels and bridges, many of which are over a century old, like the Virginia Avenue Tunnel.

In 2008, just before the financial crisis, CSX launched a multistate public-private partnership to do the work, called the National Gateway initiative. The $842 million program includes sixty-one double-stack clearance projects and the construction of six intermodal terminals across Ohio, Pennsylvania, Maryland, North Carolina, Virginia, West Virginia, and the District of Columbia.[21] Although the financial crisis hit CSX in the pocketbook as its shipping customers cut back because of falling consumer demand, Michael Ward persisted. CSX found support from federal, state, and local government agencies eager to create jobs and economic growth, and CSX also contributed heavily from its own coffers.

Investing in infrastructure in order to drive economic recovery has happened before in the National Gateway. In December 1933, in the midst of the Great Depression, the Public Works Administration gave the Pennsylvania Railroad a $77 million loan to create 45 million "man-hours of direct employment," in part on a project to improve the same Virginia Avenue Tunnel. Now it needed help again. In 2009, the economic stimulus package signed into law in February, at the height of the Great Recession, included $1.5 billion in competitive Transportation Investment Generating Economic Recovery (TIGER) grants. The ability to create 50,000 jobs over thirty years—including 10,000 construction jobs in the near term—made the National Gateway a promising candidate for funding. In February 2010, the federal government awarded CSX and the State of Ohio a $98 million grant. Seeing the potential economic benefits of modernized infrastructure, state governments in the National Gateway partnership committed another $180 million. CSX itself contributed $575 million toward the program.[22]

There's another imperative for CSX and the National Gateway: competition. CSX's principal rail competitor in the East, Norfolk Southern, has been pursuing a similar strategy in the Louisiana–New Jersey Crescent Corridor, as I mentioned earlier, and has already completed and opened the $290 million Heartland Corridor, a public-private partnership that cleared twenty-nine tunnels in Virginia and West Virginia for double-stack operations and reduced East Coast–Midwest trip times by an entire day.

Dozens of infrastructure improvement projects have been completed as part of the National Gateway initiative. The effort is on track to be finished in time for the significantly increased traffic through East Coast ports that is expected to come with the opening of an expanded Panama Canal in 2015.

The Virginia Avenue Tunnel is next. As I write, construction is ready to proceed, as long as it overcomes community controversies.

The Federal Highway Administration and the District (of Columbia) DOT have issued an environmental impact statement, which includes community concerns, addressing objections raised by nearby Navy Yard residents. The neighborhood includes less-affluent parts of Washington, was once home to an "adult entertainment" area, and has been under redevelopment since the 1990s. It is now the site for a U.S. Department of Transportation office complex and the nearby Nationals Park stadium for the Major League Baseball team, making the area of interest to all sports-loving Washingtonians.

CSX will not focus only on the rail tunnel. The company promises enhancements to the new streetscape to complement other development and raise standards. CSX will straighten roads, improve park access for the wheelchair dependent, build a continuous bike path, enhance landscaping, and install improved street lighting, traffic signals, and crosswalks.

That's the other thing to remember about the right kinds of strategic infrastructure investments. They not only produce jobs for today and economic opportunity for tomorrow; they can also enhance quality of life for surrounding communities. And when the demonstrations are in the nation's capital, perhaps the gridlocked Congress will notice. Its members certainly notice when their constituencies are passengers.

THE PROBLEM FOR PASSENGERS

One catastrophe sums up the challenges for passenger rail and the significance of aging infrastructure for everyone in the neighborhood, whether or not he or she uses rail. On Friday, May 17, 2013, at 6:01 p.m., Metro-North Railroad 1548, a commuter train traveling east from New York City to New Haven, derailed, partially obstructing the adjacent track. Twenty seconds later, it was struck by a westbound Metro-North train. The conductor of the second train had applied the emergency brake, and both trains were relatively new, sturdy models. Nevertheless, the 23-mile-per-hour impact of the westbound train shattered the insides of the train cars and ripped off their steel shells "like ribbons of clothing," an observer said. Seventy-three passengers, two engineers, and a conductor were injured. The direct cost of the damage to the Metro-North system was $18 million.[23]

The significant human and financial harm was magnified by the location of the crash. The four-track-wide stretch in Bridgeport, Connecticut, was one of the busiest in the United States, being used both by Metro-North for commuter services and by Amtrak for inter-city passenger transportation; scores of trains crossed the site every day. At the time of the accident, two of the four tracks were closed for scheduled maintenance of the electrical system. When the crash destroyed the other two tracks, virtually all rail services eastbound from New York were suspended, including highly trafficked routes to New Haven, Providence, and Boston. Locomotives between Boston and New York were stranded, creating ripple effects that disrupted travel along the Eastern Seaboard, where the Boston–Washington Amtrak route alone carries about 700,000 people a day.

The crisis spilled over, as the 30,000 people in Connecticut who normally commute by train took to the roads instead. One commuter who spent ninety minutes on what is usually a twenty-minute drive to the Bridgeport train station complained about the nightmare of just getting into the city that day. Urging residents to stay home, or at least to carpool, Governor Dannel Malloy of Connecticut warned that if all the commuters were to get on highways in single-occupancy cars, it would turn the roads into parking lots. Bridgeport mayor Bill Finch remarked that his city's pipeline to New York City was going to be shut down for some time, costing the region a great deal of money, not just for the repairs but also because of the lost wages and reduced economic activity.

Two days before the accident, a routine visual inspection had found that a rail joint near the scene of the accident had inadequate supporting ballast and indications of vertical movement, the

National Transportation Safety Board reported. That single broken rail became the most immediate suspect, although it was just one of many trouble spots resulting from years of deferred maintenance of components dating to the late 1800s. The catenary wiring that supplied electricity overhead is 110 years old. (It was the first mainline electrification in the world.) Upgrades have been under construction for 20 years.

A mere 2,000 feet of track was ripped up, less than half a mile, and it is possible to redo a mile of track a day, so repair should have been swift. In fact, it took nearly six days before Metro-North and Amtrak services were restored. Fragmentation was one cause of the delay; numerous local jurisdictions had to be involved, as well as state and national entities and the railroads. Moreover, track in the Northeast is used and owned by a complex web of stakeholders, including commuter rail agencies, freight train operators, and Amtrak.[24] Then there was government bureaucracy, which delayed approvals, and union work rules, which delayed crew deployment.

U.S. Senator Chris Murphy of Connecticut said that $4.6 billion would be needed in the next two decades for rail infrastructure in Connecticut alone to be in good repair. He and others in Congress called for the creation of a rail trust fund—modeled on the Highway Trust Fund—that would sell Treasury bonds to support train and track investments. Soon the outcry died down, although in June 2014 Murphy cosponsored a bill in the Senate with Republican Bob Corker of Tennessee to raise the gasoline tax for two years to pay for transportation infrastructure improvements. Nothing happened.

The Metro-North crash illustrates many of the problems with U.S. passenger rail services, which remain woefully underdeveloped, analysts say—especially when compared with most other wealthy OECD countries. Or compared with freight rail.

AMTRAK AND THE PASSENGER TRAIN PREDICAMENT

Although the United States has the world's largest national rail network and one of its largest populations, America ranks a middling seventeenth in passenger-miles of services provided each year. This is barely above Iran and well below both emerging markets like China and India and low-density developed economies like Russia. Indeed, on a per capita basis America performs even worse. With a per capita annual ridership of less than 50 miles, it ranks thirty-second, below emerging economies like Indonesia, Kazakhstan, and Pakistan. In contrast, per capita passenger rail usage is over 750 miles in Russia and over 850 miles in France.[25] These rankings reflect the American deprioritization of passenger rail over other modes of transportation.

Since the passage of the Rail Passenger Service Act in 1970, the National Railroad Passenger Corporation, known as Amtrak, has been the only nationwide provider of passenger rail services in the United States. A wholly government-owned corporation, Amtrak provides services to over 31 million passengers across three lines of business: the Northeast Corridor, short-distance and state-supported routes, and long-distance routes. But despite increased management flexibility granted in its 1997 reauthorization, Amtrak has been plagued by internal turmoil and turnover. We spoke to many current and former Amtrak officials and board members, all off the record because of political sensitivities. Amtrak's 1997 vision of high-speed rail remains undelivered and unlikely. Even advocates for leaps forward in transportation and infrastructure are unsupportive. General Electric CEO Jeffery Immelt says, "I can tell you as a private investor, as a private company, I've had the chance to invest in high-speed rail a dozen times, and I've said no every time, because I didn't think it made sense from an economic or technical perspective."

In 2012, Amtrak recorded an operating loss of $1.2 billion on revenues of $2.9 billion; losses ranged from 39.0 percent to 49.1 percent over a five-year period. But financial performance varied. The Northeast Corridor, one of Amtrak's flagship services, accounted for 11.4 million riders in 2012 and an operating surplus. It includes the Northeast Regional route, which stretches across major population centers from Boston to Newport News, Virginia, as well as Acela Express services from Boston to Washington, D.C.—the latter a higher-speed route that aims to compete with airlines' shuttle services. Amtrak's twenty-four state-supported and short-distance routes (under 400 miles in length) carried 15.1 million passengers in 2012 and are the railroad's fastest-growing business line. Designed to be time competitive with other modes of travel, these types of services include routes like the Pacific Surfliner (San Louis Obispo to San Diego), the Vermonter (St. Albans to Washington, D.C.), and the Illinois Services (Chicago to Carbondale, Quincy, and St. Louis). In 2011, state funding of $185 million made up 31 percent of revenues for the routes, keeping losses to $166 million. Combined, the Northeast Corridor and state-supported routes achieved a positive operating balance of $47 million on revenues of $1.6 billion, according to data compiled by Robert Puentes and a Brookings Institution team.[26] So, depending on how the math is done and costs are allocated, some parts of the system can make money.

Amtrak's eighteen long-distance routes, over 400 miles, are seen by supporters as integral to the maintenance of a national passenger rail system. "If you are a passenger operator, long-distance trains keep your options open," says Marc Hoecker, a former passenger manager now working for an eastern Class I carrier. "If they go away, and you lose the stations, the tracks and, above all, the land, good luck trying to get any of it back." The long-distance routes carry millions of Amtrak passengers, but they account for all of Amtrak's losses, especially the fifteen routes over 750 miles in length. In 2011, with aggregated revenues of $518 million and costs of $1.1 billion, Amtrak lost $598 million on long-distance routes, even before capital expenditures.[27] These long-distance routes include the 2,438-mile-long Empire Builder service, from Chicago to Portland and Seattle, and the Texas Eagle, which runs from Chicago to Los Angeles via Dallas and San Antonio. They provide a baseline of connectivity among small towns and between rural areas and major cities in the twenty-three states where they are the only passenger rail services available.

If the money-losing long-distance trains did not give lightly populated places like Montana its only good public transportation, federal funding for the Northeast Corridor would be politically difficult,
a former Amtrak executive said.

Commuter rail services account for approximately $1.8 billion in
rail industry revenues. In 2012, they provided over 466 million trips,
a 13 percent increase from 412 million trips in 2002 and a 43 percent
increase from 326 million trips in 1992. Commuter rail services in
the United States are operated by twenty-seven different entities,
the vast majority of which are government-controlled public transit agencies focused on specific regions and primarily subways. For
example, commuter rail services in the New York City area—the
nation's busiest in 2013, with 14.8 million rides per month—are delivered by the Metropolitan Transit Authority, an independent agency
with board members appointed by the state governor, New York
City's mayor, and suburban leaders. Similar governance structures
are found in the country's four other major commuter rail providers:
Newark's New Jersey Transit (6.2 million monthly riders), Chicago's
Metra (5.7 million), Boston's MBTA (3 million), and Philadelphia's
SEPTA (2.8 million). Most others have as few as 40–50,000 riders
per month.[28]

Like those of all publicly operated passenger rail systems globally,
commuter rail revenues cover under 40 percent of expenses; public
transit agencies in general lose about $36 billion a year across all
services. However, whether owing to their contribution to economic
development or to a view that they are public services, commuter rail
services are heavily subsidized by federal, state, and local governments. Just the federal government's contribution to public transit
agencies in 2011 was nearly twenty times Amtrak's operating loss in
the same year.[29]

The poor financial performance of passenger rail services has
led to underinvestment. Unlike freight railroads, Amtrak has no
profits to reinvest in maintenance, much less in ambitious capital
projects that would make American passenger rail services truly

globally competitive. Although some insiders say that Congress set Amtrak up for failure, some current leaders in Congress view Amtrak's continuing losses as evidence of poor management, and argue that they should not throw good money after bad in the form of public subsidies. "Amtrak is the popular punching bag for Congress," the industry insider said. He explained that Amtrak's poor revenue should not be evaluated as a reflection of poor management, because no publicly operated passenger rail system in the world has a profitable revenue stream. The result has been an infrastructure deficit almost unparalleled in the world. According to the Organization for Economic Cooperation and Development (OECD), the United States must invest an additional $230 billion between 2015 and 2030 to restore its rail infrastructure to a level of global competitiveness. The United Kingdom, in contrast, is $48 billion behind and Canada's infrastructure deficit adds up to a relatively small $28 billion between 2015 and 2030.[30] If past is precedent, this funding gap is unlikely to be filled by government. Rail's share of U.S. federal transportation funding *peaked* in 1976 at 4.95 percent and has since declined to 1.02 percent. This contrasts with highways' 80 percent share of transportation spending; aviation industry subsidies round out the remaining 19 percent.[31] Do we need any more evidence that trains are the neglected stepchildren of America?

Capital investment in commuter rail infrastructure declined to under 15 percent of public transit investments in 2011. Of all public transit modes, commuter rail services have the highest share of antiquated infrastructure; 24 percent of commuter railcars were manufactured before 1980. Mass transit agencies have also been slow to increase the efficiency of commuter rail infrastructure. Although the share of buses powered by diesel fuel decreased from 81 percent to 64 percent between 2006 and 2011, the share of locomotives powered by diesel declined by only half a percentage point during the same time period.[32] (More about public transit later, as a central aspect of rethinking cities.)

The lack of investment in American passenger rail services has led to comparatively poor performance. Speed is one metric: a "high-speed" trip from New York to Washington on the Acela Express takes two hours and forty-two minutes; it would take only one hour and fifty-five minutes on a European or Asian high-speed train. Similarly, the Boston–Washington Acela journey currently takes six and a half hours compared with the potential to do it in just over four hours with technology that is commonly used abroad. It's little wonder, then, that Amtrak's customer satisfaction index has barely improved over the past few years. Or that Amtrak insiders are so hesitant to talk on the record.

Then there's the connection challenge. Unlike freight rail, which maintains tight connections with trucking companies and ports, passenger rail doesn't play well with other modes of transportation. This further decreases efficiency, raises costs, and makes passenger rail less attractive to passengers. An industry insider said, "According to the MBTA (Massachusetts Bay Transportation Authority, which runs subways, buses, and commuter rail), there is no Amtrak. None of the timetables are integrated. All the agencies act locally. For example, no one can understand low use of T. F. Green Airport in Providence. MBTA uses it as a commuter station. A bunch of trains pull out in the morning heading to Boston, but nothing pulls in. You can't get to the airport from any point north of Providence. The State of Connecticut runs up the Northeast Corridor, which is a natural connection, but nothing goes there. Amtrak trains slow down at T. F. Green but don't stop. There is no fare integration. Fares are all different, all local. Google makes an attempt. Everyone talks about standardizing, but they don't."

Intramodal and intermodal connectivity issues cut across all forms of transportation. Mobility does not stop at local and state borders, yet funds are often allocated to states without regard for how they will work together. Hypothetically, if Ohio were to build a 200-mile-per-hour rail capacity, does that stop at the Indiana border? And if Indiana decides to restore a third track to get freight trains off two main lines, can Illinois, which is out of money, remove the benefits by doing nothing? Improvement requires collaborative innovation.

The problem of underinvestment exacerbates the innovation problem. America is also behind in train manufacturing and servicing. The countries with the most competitive rail manufacturing industry make a large, steady stream of investments in rail and public transit, which creates substantial domestic markets and demand for innovation.[33] Although America's freight rail service industry is robust, freight trains run relatively slowly and do not require the same technical sophistication as modern passenger trains. The lack of sustained, ambitious investments in advanced passenger services has led to comparatively low demand for innovations in speed and performance.

We should all care that America's position in the global train manufacturing and servicing industry is just middling and want it to be better. Of the top ten train manufacturers in the world by 2010 revenues, four are European and two are Chinese. General Electric Transportation is the only American firm on the list, ranking tenth, with revenues of less than a fifth of the market leader, China's CSR Corporation. Of approximately twenty top-tier original equipment manufacturers serving the U.S. domestic market, thirteen are foreign firms. But most of the foreign-owned firms serve broad swaths of the market, while American firms tend to be more narrowly focused. For example, Alstom (France) provides train sets for intercity passenger rail, high-speed rail, regional rail, metro rail; Bombardier (Canada) supplies each of those types of trains, plus light rail systems. Bombardier and Kawasaki (Japan) together account for more than half of the new railcars sold in the United States. Indeed, in mid-2010 (the most recent comprehensive market analysis), only six firms supplied 94 percent of the American railcar market. In contrast, three of the U.S. firms—EMD, GE Transportation, and Motive Power—manufacture only locomotives, for intercity and regional passenger rail; three other manufacturers—Brookville, Gomarco, and United Streetcar—were only in the streetcar industry; and US Railcar makes cars only for intercity and regional passenger rail.[34]

"Buy America" rules require federally funded rail infrastructure to be assembled in the United States and to be composed of at least 60 percent domestic content. To comply, foreign firms have set up assembly plants in the United States. An estimated 159 suppliers, of which 135 have U.S. headquarters, provide propulsion components like engines and suspensions, electronic systems like driving control systems and communication systems, and body and interior components ranging from heating and air-conditioning systems to doors and bathrooms. Some belong to large conglomerates; others are small companies with fewer than one hundred employees. The relative importance of the rail segment for these companies also varies, accounting for anywhere from 20 percent to 90 percent of revenues. But despite the law's intent, the bulk of design and other high-value work that foreign companies undertake is done overseas.[35]

In America, it seems, rail hasn't been important enough to anyone to invest sufficiently in repair or renewal, let alone innovate to reinvent rail transportation for future needs. That's why all Americans should care. It's a self-reinforcing cycle. A strong domestic industry serving a demanding domestic market encourages competition to reach those customers, which in turn encourages innovation. Of course, without innovation, customers seek alternatives, which weakens demand and moves investment elsewhere. Finding new approaches is essential.

HIGH(ER) SPEED AT LOW COST: HOPE FOR AMTRAK

Here's the million-dollar, or maybe the billion-dollar, question. Can the same kind of leadership and innovation that is removing community pain points and readying freight rail for the future work for passenger rail?

High-speed rail has been a pipe dream for decades. Only a few wild dreamers can imagine that the United States will invest in bullet trains or maglev systems. Entrepreneur Elon Musk, Tesla electric car founder, has envisioned a futuristic high-speed system called the hyperloop, consisting of solar-powered pressure tubes that could propel trains ten times faster than high-speed rail, but it is nowhere near reality. (*Star Trek* fans like me might as well wait for teleportation.) Meanwhile, trainloads of consultants, including McKinsey's global infrastructure practice leaders, have argued that infrastructure problems can be solved at low cost by leveraging existing assets, using technology, and improving project management.

Lo and behold, beleaguered Amtrak itself is proving that existing passenger rail can be speeded up enough to compete with automobiles. The Keystone Corridor project demonstrates the potential of relatively low-cost incremental improvements that do not radically change the underlying design but rather build off it with innovations that can increase performance sufficiently, if not dramatically.

For nearly two centuries, the Keystone Corridor rail line running from Philadelphia to Pittsburgh has played an important role in the development of the region and the mobility of residents. Opened in the 1830s by the Main Line of Public Works—a hybrid canal and rail system developed by the Commonwealth of Pennsylvania—the 349-mile route was the principal artery of the Pennsylvania Railroad system. In the early twentieth century, the railroad grew to become America's largest, and it made investments in its Philadelphia–Harrisburg (the state capital) route a priority, electrifying the line between the 1910s and 1930s. Over time, the Keystone Corridor proved instrumental in the creation of some of America's first suburbs; wealthy communities such as Bryn Mawr, Haverford, and Radnor took the names of stations along the route and are still collectively referred to as the "Main Line" towns.

By 2000, several privatizations, bankruptcies, and nationalizations later, the Keystone Corridor was used by multiple rail operators. Amtrak provided its Keystone Service from Harrisburg to New York City and its Pennsylvanian route from Pittsburgh to New York. The Southeastern Pennsylvania Transportation Authority (SEPTA) operated a 35-mile commuter rail service from Thorndale to Philadelphia. Norfolk Southern leveraged the corridor for its overnight freight services connecting to the Crescent Corridor.

The corridor's infrastructure was in the worst condition since the Great Depression. In the 1960s, facing a declining passenger business, the Pennsylvania Railroad had actually *removed* two sets of track. When the railroad finally declared bankruptcy in 1970, responsibility for the Keystone Corridor was split between the newly created, government-owned Amtrak and Conrail corporations (with Conrail later reprivatized under CSX and Norfolk Southern). Decades of underinvestment led to "a gradual decline in operating speeds, travel times, ride quality and reliability," a senior Amtrak executive said.[36] The 105-mile Philadelphia–Harrisburg trip often took over two hours, barely an improvement over travel times in the early 1900s.

In 2000, recognizing the strategic importance of the Keystone Corridor, officials from Amtrak and the Pennsylvania Department of Transportation (PennDOT) began discussing infrastructure improvements and produced a weighty planning document for the Federal Railway Administration. Two years later, they announced the goal of making the Philadelphia–Pittsburgh route "auto-competitive" by 2015—meaning that passenger trains could steal travel share from cars. The critical element would be increasing train speeds from their current limits of 90 miles per hour—70 in some particularly dilapidated stretches—to 110 miles per hour across the entire route. This would decrease the Philadelphia–Pittsburgh travel time from 120 minutes or longer to just 90 minutes for express trains, or 105 minutes for locals.

This is a far cry from high-speed rail in China, Japan, or Europe, but it would prove the point that passenger rail can compete effectively with cars, and at relatively modest cost. It would inch closer to European TGV speeds, without building the completely new, expensive tracks that bullet or maglev trains require.

The engineering relied on a comprehensive set of incremental improvements across every aspect of the infrastructure, rather than a single expensive new breakthrough. More than 85 miles of decades-old wood ties were replaced by concrete, and 105 miles of new "continuous welded rail" track were deployed. An antiquated, manual, and only one-way signaling system was replaced with modern communications equipment that allows for two-way operation. Forty miles of new catenary overhead wires and thirty-six new circuit breakers were installed to overhaul the electric power system, which dated back to 1915; thirty-eight bridges were upgraded. The overall capital cost of the Keystone Corridor was $145.5 million.[37]

The collaborators worked together in innovative ways, at least as bureaucracies go. The Keystone Corridor project marked a long-term commitment rather than the usual year-at-a-time subsidy from the state to Amtrak. PennDOT executive Joe Daversa remarked, "All that [annual subsidy] amount does is buy some supplemental service," whereas a comprehensive five-year capital improvement plan would allow the service to run "the way the line was originally designed to run."[38] Departing from precedent, the capital budget for the project was split—evenly—between Amtrak and PennDOT. Instead of approaching the Keystone Corridor revitalization as a multibillion-dollar megaproject, state and federal leaders looked for the most cost-effective way to generate just enough performance improvement to make rail transportation competitive with roads, so that taking the train is a welcome alternative to driving.

Faster services went live in 2006. By 2008, ridership had grown threefold over that in 1998, to approximately 1.2 million people, a dramatic increase from 400,000 in 1998—a level of growth that far outpaced that of the overall Amtrak system. The Keystone Service and Pennsylvanian are two of Amtrak's fastest-growing routes; ridership increased by 5.2 percent and 4.3 percent, respectively, in the first half of 2013 alone. Revenue per passenger mile increased and state subsidy per passenger mile decreased. In 2011, PennDOT and Amtrak received a total of $66.4 million in grants from the federal Department of Transportation through the American Recovery and Reinvestment Act's High Speed Rail program to eliminate the route's three remaining highway crossings, allowing trains to travel at speeds of up to 125 miles per hour.[39] That will be even better.

Contrast the $1.34 million spent per mile on the Keystone upgrade to the well over $12 million per mile estimate for the California High Speed Rail project. I like those Keystone economics. So will the *Wall Street Journal*-reading business executives now whizzing to and from their Pennsylvania destinations faster.

Compared with the magnitude of the need, and the many years that the problems have been piling up, this effort can seem small. But small wins demonstrate what's possible, inspire others to act, and can accumulate to make a big difference.

LEADERSHIP FOR THE FUTURE: RAIL BACK ON TRACK

America can't leave rail transportation stuck in the last century or two, grappling with a legacy of past neglect and a failure to find sufficient investment for the future. Rail is too important as a long-distance and commuter alternative for moving goods and people. The potential for quintuple wins is clear: potential safety compared with highways, cost-effectiveness for passengers and shippers, productivity thanks to reduced road traffic congestion, lowered carbon emissions for a greener environment, and economic development benefits. But each of these is jeopardized owing to infrastructure badly in need of repair, let alone waiting for renewal and reinvention.

The first of the three R's, repair, is essential, and it must be a national priority with incentives for the states and private industry —the freight railroads—to do more of it. The National Gateway initiative and the Crescent and Heartland Corridors are collaborative models that should be duplicated everywhere, not just because of the direct repairs and upgrades, such as raising tunnel heights, but also because rail-oriented projects can end up enhancing many other things in the vicinity, including roads and parks for bikers and pedestrians, as we see in Chicago's CREATE projects. Railroad companies should be encouraged to be strong community partners, addressing community needs, as CSX does, or working with local ministers concerned about traffic fatalities, as Norfolk Southern did. Railroads pass through towns, but their leaders should stop long enough to become part of the community-building strategies of every town and region. National standards could ensure that repair is always associated with renewal—for example, the no-brainer of replacing manual switching with automation, adding sensors for safety, or finding new materials for wooden ties and wiring.

A broad capacity constraint will affect freight transportation services in the Mid-Atlantic region over the coming decades. By 2050, the U.S. population is expected to grow to over 400 million, and the corresponding demand for freight will double from 15 billion to 30 billion tons per year. Ports in Baltimore, Hampton Roads, Virginia, Wilmington, North Carolina, and elsewhere in the Southeast have been upgraded to accommodate rapidly increasing volumes of shipping containers. However, little has changed in the infrastructure used to deliver goods from these ports to and from the interior of the United States. As demand grows, consumers are likely to face long waits to buy imported goods, and American manufacturers will be delayed in bringing their products to global markets. If all of this new freight traffic were added to the region's highways, instead of using rail, there would be crippling delays for commuters and 20 million tons of new CO_2 emissions from trucks.[40]

America must move to modernize rail. Rail transportation offers viable alternatives for a future likely to be characterized by greater urban density, a demand for mobility not dependent on cars, and a continuing need to reduce air pollution and traffic congestion. Serving the poor in cities, who rely on public transit, as well as the disconnected in rural areas, remains an unsolved problem. The Panama Canal expansion promises to bring bigger ships with larger cargoes to U.S. ports and, along with them, more business opportunities, but also greater demand on limited capacity. All these trends call for rethinking and reinventing rail transportation, while improving the infrastructure that supports it.

Innovation can help the United States move toward exciting concepts like bullet trains and maglevs without trying to duplicate these costly futuristic systems, which would require entirely new infrastructure anyway. That kind of innovation might take place in new sites and with private-sector funding; Texas Central Railway is a private company planning to bring bullet trains based on Japan's Shinkansen system to the corridor between Dallas and Houston, with speeds over 200 miles per hour, making the trip in ninety minutes. All Aboard Florida wants to connect Miami and Orlando with service like the Acela, rather than truly high speed. Northeast Maglev is a much grander vision: a maglev train for the entire Northeast Corridor, complete with the new tracks that a maglev requires, creating the fastest train in the world. Total cost would be a gargantuan $100 billion. Its first leg, between Baltimore and Washington, D.C., is a $10 billion, ten-year effort that has received several billion dollars in funding from the Japanese government, that features a prestigious advisory board of former governors, cabinet officials, and business leaders, chaired by former U.S. Senate Majority Leader Thomas Daschle, and that has raised numerous skeptical eyebrows. Maybe U.S. technological innovation could even leap beyond to new concepts. Tesla CEO Elon Musk and his hyperloop idea make clear that private-sector inventors are already imagining new possibilities, even if, like the hyperloop, they remain dreams.

It is important to encourage entrepreneurs and innovators, but also to promote innovation in the repair and renewal process. The Keystone Corridor project has already increased rail speed; a next round could seek manufacturers who can work on further enhancements. More people demanding rail services, including commuter rail and public transit in cities, can create a favorable investment climate for new ventures to offer new visions. One clear need is to situate rail in a broader transportation system that includes connections across modes of transportation, so that we could indeed get one ticket and bar code for a trip by plane and train—assuming we figure out how to make the physical connections, which should also be a priority. And potential visionaries need places to learn the industry. Many children and even adults own train sets. Schools, however, teach next to nothing about railroads. Community colleges could take this on.

We can overcome funding constraints by encouraging funding consortia: multiple funding sources, using multiple methods, spreading risk while ensuring broad benefits. Public-private partnerships can encourage collaboration and coordination across sectors, jurisdictions, and modes of transportation, toward integration of pieces into seamlessly connected sources of mobility. With leadership from both public and private sectors, coalitions are possible; the involvement of business gives cover to politicians wary of asking the public to fund long-term investment. Government officials can speed up the decision-making process when they want to—or when the private sector pushes, dangling the promise of jobs.

All this would lay the track for a future-ready industry, in which companies that otherwise compete and jurisdictions that operate independently could put aside their differences in the interest of building the platform for the industry to flourish and the nation to be better served. Public visibility for rail in all of its forms, both freight and passenger, long-distance and short transit within cities, would make clear its environmental and economic contributions. Public support would attract private investors.

We might not have the glitziest and most expensive new trains, but there are many other ways that the United States could again become an innovator for our communities and the world. Fortunately, freight rail is carrying its share of the load.

3

UP IN THE AIR

How can something meant to be so fast often feel so slow? Airplane passengers just want to get there. But as delays become more endemic, they're often stuck. And flight delays are a metaphor for progress delayed.

Air transportation is a classic U.S. strength that is showing its age and needs rejuvenation. Air made its ascent as a dominant transportation mode after World War II while rail declined, and then flew high. English became the universal global air transportation language, and U.S-trained pilots were in significant demand. The United States led the world with the best manufacturers, best in-flight operating systems, biggest market with the most domestic routes, technology innovations for passengers and in the cockpit, and competition that drove prices down.

The past few decades have seen a steady descent from great heights, and that costs the economy, the airline industry, passengers, and the environment. There is congestion on the ground, congestion in the air, and reliance on ground-to-air air traffic control systems designed decades ago, modernizing steadily, yes, but very slowly. Regulators with outdated premises can be at odds with tech entrepreneurs, delaying the use of world-class technology innovation. After numerous bankruptcies, a few U.S. airlines have emerged stronger than ever, but the industry in general still falls behind international competitors. Aging airports get low marks, and priorities for modernizing them do not always fit transportation strategies. Fuel, already costly, is often wasted, and carbon emissions are alarmingly high. Weather volatility poses an increasing challenge, with weather not so severe as to shut everything down but bad enough to shake everything up. Industry leaders develop agendas, but there is no national vehicle to address the challenges together, and the Federal Aviation Administration is captive to the congressional budget process.

A convenience for long-distance intercity passengers, air transportation is also a catalyst for commerce and a boost to the American economy, moving over $562 billion of freight and nearly 800 million passengers by air per year. The U.S. civil aviation industry creates $1.3 trillion in economic activity, supports 10.2 million jobs, and represents 5.2 percent of U.S. GDP each year.[1] As the leading exporter of engines and aircraft equipment, the U.S. civilian aviation manufacturing industry has contributed a $75 billion export surplus, improving the overall U.S. trade balance.

The United States ranks high on some global comparisons, garnering a number 2 on the 2013 World Economic Forum Global Competitiveness Report for air transportation (Canada is number 1), boosted by first-place rankings in available seat-miles and the number of operating airlines.[2] Americans flying domestically have perhaps more choices than residents of any other country. North America is the largest aviation market in the world, with the Northeast Corridor from Washington to Boston among the busiest.

For all the flying we Americans do, you'd think we love it. Instead, we often hate every mile of the trip. There is deep rage about limited leg room, and fights break out over whether the passenger in front may recline her seat. Then hit one month of bad weather in the country, and delays and complaints rise together. In May 2014, when on-time arrivals were down, 8,300 flights were canceled because of storms, about half of the month's cancellations, and passengers filed 1,010 complaints with the government, up from 730 a year earlier.

The ready availability of low-cost seats, however squeezed, has overwhelmed use of intercity passenger rail. Planes, along with cars, were the lucky beneficiaries of 1950s policies. But the U.S. system is also characterized by crowded skies; price competition among airlines and low profitability; competition among airports, leading to congestion in some places and wasted capacity in others; concerns about modernization of ground facilities; a dearth of intermodal links such as air-to-train connections; high fuel costs and utilization, and air pollution; and dependence on outdated intergovernmental agreements for access to international markets.

The United States ranks a mediocre number 30 in the world for quality of air transport infrastructure as measured by a survey of executives, a downright low number 127 in ticket taxes and airport charges (meaning that they're too high), and an even lower number 131 in carbon dioxide emissions per capita (way too high), because more flying equals more pollution. And there are worries ahead. The American Society of Civil Engineers, which grades America's infrastructure, argues that a failure to invest in aviation could represent an estimated cumulative loss of $313 billion by 2020 and $1.52 trillion by 2040 and potentially lead to over 350,000 fewer jobs by 2020.[3]

That's a lot to worry about while being crammed into small spaces. But rest assured that the pilots who fly the planes also worry about the pain points and bottlenecks.

Captain Brian Will, a top pilot at American Airlines, is waiting not too patiently for improvement. He can tell you why he wants change. "It's really simple. When things go bad around the system, things start backing up, we start getting delays, and I have 150 or 250 people strapped to my back. You're looking at the gas gauge, and if you're not exactly sure, and you're 150 miles from where you've got to land, well, that's not a good place to be. There are things we can do today that are far more efficient and give me a lot more clarity," he says, adding that they aren't being implemented fast enough. We were talking in May 2014 at AA's System Operations Control Center in Fort Worth, Texas, a few miles from the Dallas–Fort Worth airport, in a conference room next to a balcony overlooking the nerve center of the airline, the flight dispatch floor, where every AA flight in the world is launched and tracked.

Confident and articulate, Will could be a movie version of the pilot-hero. After ten years in the U.S. Air Force as a fighter pilot flying the formidable F-15, he joined AA in 1989 with degrees in physics and aeronautical sciences. Besides captaining the most advanced aircraft, he serves as AA's director of Airspace Optimization and Aircraft Technology and is a leading spokesperson for the importance of a new generation of technology, familiarly called NextGen. He is among those working to bring air transportation into the era of smartphones, iPads, mobile communications, cloud computing, and real-time weather forecasts.

Will's vision: innovation can produce multiple wins for everyone. "If we offer a more efficient product, we can cut prices. If we reduce flight time and distance, that cuts carbon emissions. We can cut noise by being at idle power for long portions of the arrival. The communities like that; the customers at our airports that we serve like that; air traffic control that releases the bird like it," he said. Then, recognizing that he was sounding like a pitchman, he explained, "I've heard people say, 'Come on Brian, you're selling something here,' but really I'm not. Because when I'm up in the air, and things are not going well, I'm strapped in the jet too. There's really no downside. Everybody wins—us, the passengers, environment, regulators. That's why I'd like to see NextGen change succeed."

Will isn't the only one up in the air, and his name is providential. The United States as a nation is stalled in midair, waiting to see which way the winds are blowing, and whether we will have the will to improve. How did we arrive at this destination? Are we stuck here, or can we move to a better place? The challenges faced by the air transportation system have roots in the past.

FLYING HIGH? HOW THE U.S. AIR
TRANSPORTATION INDUSTRY TOOK OFF

In 1903, the Wright brothers made their first successful flight, as the world knows, and are credited with inventing the airplane. On January 1, 1914, the first commercial flight traveled from St. Petersburg to Tampa, Florida, with one paying passenger on what was called an airboat line. In 1915, a national advisory council was established to focus on aircraft—the first federal agency following pretty quickly after the first flight. It was the precursor to NASA, the National Aeronautics and Space Administration, which was created in 1958, a year after the Soviet Union launched *Sputnik*, with a mission to help close a technology gap in the air and outer space.

Prior to World War II, the nascent industry suffered from poor safety and many accidents. In 1940, President Franklin Roosevelt created the Civil Aeronautics Administration (CAA) and the Civil Aeronautics Board (CAB) to develop safety standards and build public confidence. The CAA managed the overall air traffic control system, while the CAB regulated airlines, which were privately owned, unlike government-owned flag carriers in most other countries. The war and the development of standards paved the way for the expansion of this mode of transportation after the war, when airlines could take advantage of new aviation technologies developed for the U.S. military and employ already trained wartime pilots who returned home.

Those were the glory days. They are still a subject of nostalgia. The Wings Club in Midtown Manhattan continues to be used by airline executives in the New York City area. It was the recruitment center for the now defunct Pan American Airways, in the former Pan Am Building. My team met with JetBlue CEO Dave Barger and VP for airports Ian Deason there, even though JetBlue's headquarters is near JFK Airport. The club is history infused, replete with photos of Presidents Harry Truman (1945–53) and Dwight Eisenhower (1953–61), whose policies jump-started post–World War II growth in the air.

Between the 1950s and 1980s, passenger air travel in the United

States grew by an average of 10 percent per year, and so did the
number of independent airlines. Growth and maturation led to a
reexamination of the role and intensity of government regulation.
The Federal Aviation Act of 1958 eliminated the CAA and established the FAA, first called the Federal Aviation Agency and later,
in 1966, the Federal Aviation Administration when it was moved
into the U.S. Department of Transportation. The FAA is the central authority for all safety-related matters, and it administers
the national air traffic control system, in a generally productive
partnership with the airlines. The 1978 Airline Deregulation Act
phased out the CAB by 1984, enabling all carriers to freely enter
new markets and routes. The legislation removed restrictions on
fares, routes, and frequency of flights.[4]

Interline ticketing, to permit connections and exchanges across
networks of passenger carriers, was a major advance, requiring both
technology and cooperation among airlines—and there were always
some holdouts. In the 1960s, in collaboration with IBM, American Airlines developed the first computerized reservation system,
SABRE (now an independent company), which gave AA a temporary
competitive advantage; European airlines followed with a reserva-
tions consortium. AA also invented the first frequent-flier customer
loyalty program, which soon became the industry norm.

In the 1970s and 1980s, deregulation encouraged domestic competition. International competition also increased with privatization of European national airlines that had formerly charged high
prices, changed slowly, and were not passenger-friendly. Their rapid
transformation began to threaten U.S. airlines on lucrative international routes. SAS, a partnership of Scandinavian nations based in
Copenhagen, began to set new customer service standards, focus on
full-fare-paying business customers (who were and are still the most
profitable), and target the U.S. market. Newly privatized British Airways styled itself as "truly global" and made London's Heathrow Airport a worldwide hub.

Independent air cargo services grew in this era, challenging government airmail services and competing effectively with the U.S. Postal Service. In 1971, Fred Smith founded Federal Express in Little Rock, Arkansas, moving it to Memphis in 1973, where the first overnight delivery service was initiated. He proposed the idea in a course paper while a student at Yale in the 1960s; he got a C on the paper, but he created a multibillion-dollar company, which he still chairs. In 1979, Federal Express pioneered the use of computers for the tracking of packages. By 2013, FedEx Express, a subsidiary, was the world's fourth-largest airline in fleet size and the largest in terms of freight tons flown. Fleets of planes, trucks, and sorting facilities using advanced technologies have helped FedEx, United Parcel Service (UPS), and DHL dominate global air cargo.

Shipping by air handles a relatively small percentage of international trade by volume, representing less than 10 percent, but it makes up more than 30 percent in terms of value and can contribute to passenger airlines' bottom lines. Worldwide cargo demand grew by 15.2 percent between 2009 and 2010, with international freight as a key driver, but the overwhelming majority of international trade by volume occurs via ocean vessel, which accounted for 8 billion tons before the financial crisis, compared with air cargo's much tinier 28 million tons. U.S. airports long occupied ten spots in the top thirty largest cargo airports. In 2012, U.S. cargo carriers had 12,367 million revenue ton-miles domestically (a ton-mile is one ton of revenue cargo carried for one mile) and 6,515.667 million revenue ton-miles internationally. International carriers collectively had 26,265 million revenue ton-miles. In 2012, U.S. cargo totaled 66.65 million tons.[5]

Passenger airlines utilize storage capacity on existing passenger routes for cargo, which maximizes aircraft space that would otherwise go unused and boosts revenues. This is much smaller in scale, accounting for 22 percent of air cargo collectively, and does not come close to the volume of large cargo-only carriers. But the major passenger airlines consider cargo important. For Delta, for example, cargo contributed close to $1 billion to 2012 revenue. Delta feels its advantage is speed of delivery for time-sensitive products such as agricultural commodities, pharmaceuticals, and perishable foods, because of frequent flights and a global network. In addition, Delta is a member of Sky Team Cargo, the only global cargo alliance, offering to provide seamless shipping to a large number of domestic and international destinations.

Defense spending in World War II and the subsequent Cold War reinforced U.S. strengths in advanced manufacturing and technology and made U.S. companies global leaders in aircraft and components. U.S. companies lead in aircraft engines and control systems. General Electric's aviation group is number 1, joined in the top ten by three other U.S. companies (number 3 United Technologies, number 4 Honeywell International, and number 6 Northrop Grumman Corporation).[6] For large aircraft, industry consolidation has left two legacy competitors: Boeing in the U.S. and the European consortium Airbus. But for smaller aircraft, where demand is growing, foreign manufacturers have been the only game in town. Companies such as Canada's Bombardier and Brazil's Embraer have become major players in the smaller, regional aircraft market; they split a recent large order for $3.9 billion in regional jets from AA, as part of its fleet renewal process. Regional airlines fly under the brand of the majors on shorter flights—with "short" now being defined as lasting up to four leg-numbing hours.

Aircraft can remain in use for twenty years or so, leading to wildly varying tools for pilots and experiences for passengers. Newer aircraft have been designed for greater fuel efficiency, and they take advantage of innovations. The Airbus A380 incorporates cameras, electronic flight displays, and a digital flight bag to reduce the weight of paperwork. The attention-getting star among large aircraft is Boeing's 787, the Dreamliner. The engine represents nearly a two-generation jump in technology, experts say, and improves fuel efficiency by almost 20 percent, reducing costs and improving environmental sustainability. When coupled with the use of composite materials, the Dreamliner also allows more point-to-point travel across long distances, improving accessibility and network interconnectedness. Its development was heralded as "the most daring, anticipated and scrutinized plane in commercial aviation history." The scrutinized part of that statement proved immediately true. From its launch, the Dreamliner was plagued by problems. In 2013, batteries caught fire while parked on the ground at London's Heathrow and Boston's Logan Airports, and a fire during flight forced an emergency landing in Japan. All 787s were grounded for more than three months so that the problems could be resolved. In mid-January 2014, battery problems resurfaced after gas was discovered coming out of a plane parked in Tokyo.[7] But with the world looking over Boeing's wings, and Boeing assuring the world that the Dreamliner is fixed, airlines such as AA are betting on it for long hauls of the future.

With all these strengths—the world's biggest market, multiple passenger choices, market dominance in air cargo, top-tier manufacturers, the best and latest engines and avionics—why have U.S. airlines struggled?

AIRLINES: UNCERTAINTY AND COST PRESSURE

Everything that happens to the economy hits the airlines fast and hard. Macroeconomic cycles of boom and bust, expansion and recession, reverberate immediately in the aviation industry, as business travelers decide whether to fly or to phone, and leisure travelers decide whether to forgo the luxury. The terrorist attacks of September 11, 2001, in New York and Washington created an unexpected demand shock, leading to a steep decline in passenger travel and the implementation of new security measures, including the formation of a new federal agency, the Transportation Security Administration (TSA), to replace private contractors in airports—a noteworthy reversal of direction. Rising commodity prices throughout the mid-2000s raised the price of jet fuel, the largest variable cost for airlines and about a third of their operating costs. To reduce fuel expense and ensure availability in the congested Northeast, Delta bought its own oil refinery near Philadelphia in 2012. CEO Richard Anderson says this is compatible with the fact that Delta has a number of subsidiaries supporting the airline; by the end of 2013, Delta was able to reduce fuel costs.

In addition to numerous external shocks, low-cost entrants began to shake up the industry just before deregulation enabled them to burgeon. Southwest Airlines started in 1971, the same year as Federal Express. Using a focused fleet, processes that permitted fast turnarounds, and smaller secondary airports outside major cities, it achieved numerous efficiencies of scale and high customer approval. It became a model for other low-cost carriers. Some new entrants failed, such as People Express, while others have expanded rapidly, such as JetBlue, which recently added international destinations and now has the largest number of flights out of airports such as Boston's Logan. Further downward pressure on prices has come from cyberspace. Since the rise of the Internet in the 1990s, travel websites allow instant comparisons and access to cheap seats. To maximize revenue, airlines have developed complex pricing algorithms to perfectly match prices to customers' willingness to pay and have added numerous charges for anything other than a seat and landing. For most of the major airlines, there is a growing gap between the amenities and prices in first or business class and the increasing squeeze and item pricing in coach or economy class. And although all passengers on a flight share the aircraft, each one might pay a totally different price for the trip. U.S. passengers have become accustomed to the agony. Pay for extra legroom, or get cut off at the knees.

Another consequence of industry consolidation is destination consolidation. Major airlines often withdrew completely from midsized cities, leaving service to their regional affiliates. Low-cost carriers that initially focused on neglected smaller cities began to reduce service to them. Midsized cities such as Memphis, once a regional passenger hub as well as the home base for Federal Express, face vastly reduced air service and little rail to speak of, which affects jobs and regional economies.

One destination almost every airline has visited is the courtroom. Since 1990, U.S. airlines have flown into bankruptcy courts and landed a remarkable 189 bankruptcies. Only one U.S. airline, Alaska Airlines, in 2014 the sixth largest, has managed to be consistently profitable over the past two decades. The overall industry during that period has had negative average net margins of –1.0 percent,[8] meaning that it's a good industry for losing money. Once-iconic brands have disappeared, and extensive merger activity has further consolidated the industry. The top tier of long-haul full-service airlines flying both domestic and international routes has been reduced to four major "legacy" airlines, holding 85 percent of passenger capacity: American, Delta, United Continental, and Southwest.

The merger of American Airlines and US Airways, completed on December 9, 2013, reflects many of the trends in the industry. US Airways was itself the result of an earlier merger, and American was just coming out of bankruptcy, the last major airline to use bankruptcy to reorganize. The AA-US combination forms the largest airline in the world, with a fleet of 1,500 planes and more than 100,000 employees. The merger represented a culmination of forces over the past thirty years or more, which AA and US claimed required an even larger scale in order for them to reduce costs and compete. The deal was first protested by regulators and held up for antitrust reasons, but an effective PR campaign lined up public and business support for the merger, which some heralded as ushering in a new era for the industry. Certainly Brian Will and the pilots, planners, and operations staff I met in Fort Worth, Texas, at AA's flight operations center in the spring of 2014 seemed determined to create that new era.

TECHNOLOGY FOR THE AIR

Technology has always been central to aviation, especially defense-related innovations in radar, avionics, communications, and computers that help move passengers and goods through airports and the air safely.

In 1930, Cleveland's airport opened the first radio-equipped control tower to manage flight traffic; airlines scrambled to equip aircraft with radio navigation systems. By the end of the decade, two-way radio technology allowed for ground-to-air communications, and radio-based air traffic control towers were constructed around the country. Successive waves of developments improved navigation, communications, and tracking of planes, under heavy regulation.

Newer developments and NextGen potential permit operational flexibility and the use of real-time data to empower pilots and airlines to make more decisions, but the potential is difficult to realize owing to classic change management issues. To get change moving requires both innovation and collaboration. Innovation is led by courageous experimenters willing to improvise, push limits, and demonstrate potential. Collaboration across the industry, among manufacturers, airlines, and government regulators, is necessary to deploy innovations under rules that guide the whole system. Collaboration is impeded by different time horizons: technology changes quickly, while bureaucracies lumber along slowly, and coalitions take time to assemble. At least, that's been the story so far.

While airlines themselves have invested in technology, to operate in the skies and take off and land on airfields, they are dependent on air traffic control. After all, human-sighted landings can fail; in 2013, a commercial airplane landed at the wrong airport.[9] Air traffic control presents significant challenges and opportunities. Aging technological infrastructure could require the FAA to make significant investments in cybersecurity, improved flight planning, and dispatch efficiency, potentially reducing delays, fuel consumption, and flight lengths. However, federal funding constraints have delayed this process and created new challenges. Current dilemmas have their roots in over two decades of wrestling with change.

NAVIGATION BY SATELLITE:
INNOVATION AND COLLABORATION

In 1991, American Airlines bought a new fleet of Boeing 757s with fighter plane–like thrust, color TVs, and a satellite-based flight management system that included a computer in the cockpit. "We've got Bill Gates up here," a pilot recalled. An AA team wondered whether that new technology would help planes navigate through volatile weather in hard-to-navigate places where it was impossible to put a radio signal on the ground. They chose Cottonwood Pass in Eagle-Vail, Colorado, a narrow mountain pass with changeable weather and often low visibility. That year, American Airlines for the first time used onboard navigation to fly through that terrain-challenged environment—it was like threading a needle with a plane—and has been doing that for twenty-three years without a single operational error.

A follower popped up almost immediately. Alaska Airlines embraced satellite technology because it had little choice, given the State of Alaska's physical climate and terrain. Supplementing radios with new technology offered advantages in lower costs and fewer delays—two reasons the airline was profitable thirty-three out of the last thirty-nine years. The new navigation system was especially important at a critical juncture about twenty years ago, when consistently low visibility in Juneau, the state capital, made landing difficult, leading some state legislators, who were often delayed getting to Juneau, to seek moving the capital to Anchorage. "We couldn't let that happen," VP and COO Benito Minicucci recalls. "Alaska Airlines got together with Boeing, the FAA, and component manufacturers, and we got innovative. We introduced satellite-based navigation into our flight management computers. We could land in conditions that were impossible before." Pilots became proficient at flying in severe weather and tough terrain. Investments in technology enabled efficient operations and on-time performance, recently the best in the industry.

By 2009, Alaska and American were working with Delta, the FAA, General Electric, and other industry partners to successfully implement other innovations such as required navigation performance (RNP), which helps aircraft glide on cruise control along a very specific path, reducing the need to switch between cruise and acceleration. This is another step closer to the holy grail of quintuple wins. "With RNP, you fly fewer miles, use less gas, create less pollution, reduce noise, get more on-time arrivals, and increase customer satisfaction," Minicucci says.

In August 2010, Captain Brian Will piloted one of the first flights using a GE Aviation–designed RNP glide system for an instrumented flight path, landing American Airlines Flight 1916 at Bradley International Airport in Hartford, Connecticut. When he explained to the passengers what he was able to do, one expressed surprise that this hadn't happened sooner. GE had already deployed RNP procedures in Canada, China, Australia, New Zealand, and Peru. The FAA had been authorizing the capability to use GPS approach routes in about one hundred U.S. airports since 2005, as part of its plan to modernize airspace by 2025. By 2025, twenty years later?

Alaska's Minicucci praised the FAA's embrace of the technology and its creation of an RNP NextGen task force to help deploy it more broadly. That's what some critics say is a bureaucracy's way to take action: form a committee, and slow things down. In this case, though, the slow pace of change is also the result of wanting a system that tracks planes everywhere, so that air traffic controllers can know where planes are and talk to them for safety reasons, even where there are no ground-to-air radio connections. That is supposedly the Automatic Dependent Surveillance Broadcast (ADS-B) program, which the FAA calls the "backbone of the NextGen system." It uses GPS satellite signals to equip air traffic controllers and pilots with real-time, accurate information about aircraft in the air—technology that would have been able to locate the Malaysian Airlines plane that was lost in March 2014. In 2010, the FAA mandated that all aircraft broadcast information into the ADS-B system by 2020. In 2013, the FAA pilot-tested ADS-B technology in four major airports, then authorized nationwide deployment. But in September 2014, the inspector general of the U.S. Department of Transportation issued a report highly critical of ADS-B that characterized the system as too costly for the benefits and warned that the system would not work unless, among other things, individual airline operators actually bought and used the equipment.[10] The system has worked in some pilot tests, but not in others.

Ten years from mandate to full deployment seems like a very long time, especially when eighteen months is a technology generation. But aircraft are big capital investments and have long lives, so it takes time for fleets to turn over. In addition, airlines are reluctant to invest in big new systems, which can cost the airline industry billions of dollars, unless there's certainty about effective operation by the FAA. Proponents still want to demand NextGen now, because "now" is always many years ahead. And while we're at it, there are a few other elements that could be upgraded if NextGen were a priority, and aviation and telecom were connected.

AA is eager to use "optimized profile descents," the idle descents from cruise to the runway by gliding that can save fuel and noise. Newer aircraft are technically capable of making a descent by gliding while on idle, but Air Traffic Control is not technically capable of accommodating this. AA invited eight FAA staff for a demonstration flight on a new Boeing 737 direct from the factory but had to reschedule when the aircraft wasn't ready. The technology launched in June 2014 in Houston and was to be gradually rolled out nationally.

THE COMMUNICATIONS CHALLENGE

Besides navigation and the tracking of planes, NextGen is aimed at communication systems that bring aircraft and airports into the age of the smartphone and tablet.

There's an industry insider joke about the tensions between pilots and dispatchers. The pilot is strapped up in the nose of the plane in a small seat on high alert while the dispatcher sits in a comfortable lounge chair drinking a cup of coffee and reading a book or leisurely strolling around. Now pilots and dispatchers are buddies because they can more easily communicate more information in real time. Breakthroughs in cloud computing, Big Data, and tablets integrate data for real-time decision-making and also make it possible for pilots and dispatchers to see the same visuals and make decisions together.

Some days are quiet in the air, like the calm-after-the-storm day I walked through American's flight operations center. A day earlier, though, big weather action in the Midwest produced a frenzy of activity as people responsible for particular routes communicated with other routes and pilots while the storms' effects rippled through the system.

When the weather changes, flight plans given to the captain many hours before the flight are questioned and revised, sometimes often. That's when pilot–dispatcher communications must be instantaneous and focused. Brian Will calls dispatchers the extra pilot in the cockpit: "When things get bad, I'm text messaging from the cockpit to my dispatcher, 'Hey, help me out here. You find me a way out of this thing. What's the best way to get where I'm going?' They'll say, 'The weather is coming through here, there's going to be a gap, and if it closes in on you, turn here and here.' These guys are awesome." With NextGen systems, dispatchers can see which flights are impacted by weather changes, pilots can be alerted, and pilots can look at what dispatchers are seeing. Better yet, they can make smarter decisions on the basis of fuller information, gaining flexibility rather than being bound to wasteful rigidity.

"In the old world in the ops center" said Des Keany, AA's head of weather planning and host for my visit to the center, "you'd have had a text product like Teletype, and you're cutting and pasting seven to eight characters onto a screen, then looking for responses from the crew in the aircraft." Teletype in the twenty-first century? That's like sending telegraphs instead of emails. Brian Will continued, "And in the days that Des is talking about, for me to respond to a controller, say from a Super 80 aircraft, I had seven characters at a time. So if I'm trying to give him an update on something, I have seven letters to do it. 'It was good,' stop. Today the stuff is awesome. I have multiple screens, and I can just type in and hit send. It's like when my kids are texting and typing on their phones."

Nothing startling there. Sounds just like what today's mobile device users are accustomed to doing, right? Except for the fact that some of this is still experimental. Although American Airlines orders 787s and 777–300s loaded with optional software and apps, aircraft fifteen to twenty years old are still flying without advanced communication capabilities. AA spends millions of dollars per aircraft to outfit its new planes with the latest navigation and communications technology, "because we believe in the technology and the safety and efficiency that go with it," Brian Will said. Some other U.S. airlines do as well—but not enough of them.

As we talked in the conference room and on the flight ops floor about all the NextGen possibilities that tech developers envisioned, one of the flight planners commented, "Some of the things they're talking about, I don't know how we're ever going to get there. How do you get dynamic airspace?" Mark Miller from the Weather Company, who arranged the visit, added, "Well, like using the iPad in the cockpit, right?" So I asked, "Why, was that a difficult thing? Is there a story there?" Ray Howland, AA's senior manager of SOC systems planning, replied, "Oh, there's a story there!"

THE FAA VERSUS THE IPAD

It's like a faux wrestling match. In one corner, we have a big, careful, slow-moving government agency dependent on the budget process in Congress. In the opposite corner, we've got exciting, dynamic technology innovations from private-sector innovators, surrounded by eager frequent-flier fans.

On the government side, even those who've led federal agencies think the current structure impedes progress. To former FAA Administrator Jane Garvey, air traffic control is still one of the most challenging parts of aviation. When she served in the 1990s, President Clinton advocated for a quasi-government entity, separate from the FAA, to manage air traffic control and free it from bureaucracy and politics, but to her dismay, the idea died in Congress (surprise, surprise). New technologies can languish unused if rules and standards are slow to be written or caught in budgetary politics. Former U.S. Small Business Administrator Karen Mills, who served in the Obama cabinet, believes that it's hard to attract entrepreneurs or investors if there's uncertainty about standards or the timing of air traffic control reforms.

As the match begins, the FAA is trying to be nimble. Insiders credit the FAA with attempting to move at what is breakneck speed for a federal agency. In 2012, the U.S. Department of Transportation was looking for lawyers to write new rules for new technologies. But innovations don't drop into the boxes already in a bureaucracy's organization chart; iPads and the idea of using them in cockpits require a new way of thinking. For regulator to say yes to anything new requires full understanding, and that educational process can be very slow. Moreover, years of outsourcing agency work have made the redesigning of procedures slower and less efficient, even when the technology exists to do it faster. Environmental impact is invoked even when it can be demonstrated that a change will have net environmental benefits. As a matter of policy and politics, the FAA's equivalent of a universal service mandate gives disproportionate voice to resisters of change, such as airlines with aircraft that don't yet have fancy tech bells and whistles and don't want to be compelled to add them. A U.S. open-airspace policy ensures that just about anything that flies can do so. One observer likened this to allowing scooters and skateboards to use high-speed highways.

The problem isn't entirely on the government side. The wireless mobile communications world seems to have caught many in the aviation world unaware. Mobile network technology is moving faster than airline information technology (IT) departments can work with. Planes didn't have much in the way of wireless connections when the iPad appeared, so IT departments didn't have network support. This is now a major focus. And the iPad was not designed as an enterprise tool, although with Apple's new partnership with IBM, that could change fast.

The wrestling is between the time frame of entrepreneurs and corporate entrepreneurs—fast—and the time frame of bureaucracies—slow. Change under these circumstances requires the same actions that have always moved new technology: a great deal of improvisation. Lead users see the potential and want to get started before the standards are fully formed. If the formal system hasn't caught up, organizational entrepreneurs find informal ways to adapt within the rules, often by using personal resources. By getting started, they help shape the standards.

"If nothing else, pilots are creative," Brian Will observes. Pilots, planners, and dispatchers are well along in figuring out how to use personal technology—their own personal tablets and smartphones and smartphone cameras—to get closer to advanced data communications in the air, to improve flights and the quintuple wins. They never jeopardize the basics of safety. They never go too far in going around. But they're definitely going around.

Put yourselves in pilots' shoes. Say you want to discuss turbulence with dispatchers before takeoff, but don't want to use the company phone on the jet bridge for fear of upsetting nervous passengers streaming aboard who might overhear. So you whip out your personal cell phone, go into the cockpit, and call flight ops to confer. What if you're not yet configured to download weather maps on your iPad? You could use the camera on your phone to take a picture of the weather screens on the computers in the crew room before boarding, and send them to your iPad to refer to in flight. Or what if you're not yet company trained or facile on the iPad, because the tools are too new? Ask your children for help.

The first use of the iPad by pilots and airlines is simple and pretty much authorized: replacing massive amounts of paper. When filled with manuals and other paperwork, the leather cases that pilots carry could weigh, on a 737 flight, about thirty-six pounds; on a 777 flight, about forty-three pounds. One pilot reported that he couldn't carry all of his books even in a forty-three-pound bag on a 777 flight, so he'd throw some of them in with his clothes and hope he could dig them out from under his underwear in time if he needed them. Now the iPad and other tablets have replaced virtually all the paper. Charts and maps will be next to move from paper to digital files.

Flight attendants could also be empowered with information. A goal of IBM's partnership with Apple, announced in July 2014, is to equip crews with iPads powered by IBM cloud-based analytics that can get them up-to-the-minute gate and connections information to pass on to passengers. That kind of information constitutes great customer service.

This obvious, basic step isn't being taken now because of another U.S. problem outside of the FAA: a lack of mobile communications connectivity. That's amazing in this age when wireless is supposedly everywhere. In the cockpit, a tablet might not be connected to anything. To check the weather, a pilot might have to take a break to go to the crew rest area and log into in-flight wireless if it's available on the aircraft. But if it's an iPad connected to a cell phone provider, there might not be coverage and a reliable connection everywhere—in fact, it's likely that coverage will be geographically limited. That could create problems, if communication is incomplete or only part of the data is retrieved. On international flights, the problem is exacerbated by different standards. Only the new 777–300 and 787 have worldwide connectivity potential—that is, commercial Internet capability outside the United States. This is why testing is under way, and why the FAA is being cautious. And as is also typical with new technology, other changes must occur, in networks and in processes within a company, to enable the benefits. At AA, two pilots are dedicated to determining how to support the iPad and other tablets.

Everyone who flies should be grateful that the FAA requires reliable, timely communication from the air and on the ground. Currently, though, communication is a hodgepodge of old and new, unconnected. Airline staff use any form of communication that works. In airports, some people walk around with old-fashioned walkie-talkies, because parts of the airport are dead spots where cell communication can't be guaranteed. Or if the weather is bad, everyone's talking on cell phones, which overloads cell networks, and some calls can't get through. Pilots use the company landline phones at gates and jet bridges. They use old text-messaging systems on the flight deck if they can't use more-modern tools. In flight, they monitor radio frequencies and listen to other pilots describe conditions. The point is to communicate in the fastest, most efficient way.

But this raises a question about how to strengthen mobile wireless in the United States so that an integrated communication system can be fast and reliable.

Progress is in the air. In spring 2014, an FAA rule change was proposed to permit operational use of tablets. If all goes well, the new rule could be issued in mid-2015. "They've realized it's a better and safer operation," an insider said, "and so I think they're trying to find more ways to say yes." There's evidence that collaboration is increasing. Perhaps the wrestling match can turn into a cooperative game.

ON TIME TO MINIMIZE DELAYS: EMPOWERING AIRLINES TO MAKE SYSTEMWIDE DECISIONS

It's clear that the new technology empowers pilots. It can also empower airlines and help passengers.

In 2003, under a joint FAA/industry/academic collaboration, a cross-sector future concepts team hit upon the idea for CTOP—Collaborative Trajectory Options Program—to deal with airspace and airport congestion and to minimize delays. In 2010, the FAA made its implementation a priority, and it was launched in early 2014, on time and on budget, to the surprise of skeptics.

Before CTOP, airlines filed a single flight plan for every flight from point A to point B. If weather, wind, visibility, or special circumstances put an airport on ground holds, that meant that the airport could no longer land as many planes as were scheduled. The FAA would slow down the arrival rate and keep departing planes on the ground longer, because circling the airport costs more and is not as safe. Air traffic control would tell the airline the time slots for each individual flight, based on the single flight plan. After CTOP, instead of one flight plan, airlines can provide five routes and a preference map. Tim Niznik, American Airlines' director of operations systems and decision support, praises the FAA's willingness to give airlines a say in flight routes when conditions require change—a form of collaborative air traffic management.

Choices help passengers. With NextGen-enabled automated flight reallocations, airlines gain flexibility and can minimize the impact of delays on passengers. The FAA can use these preferences to inform decision-making if the preferred route is not available. That way, an airline has a say in how to respond when there is a constraint in the system. Niznik compares the system to real-time highway traffic information that allows drivers to reroute according to their own preferences, rather than being controlled by a central authority that tells them where to go.

AA was particularly active in the collaboration and reaped benefits as a first mover. Innovation stemmed from collaboration. "American invests a lot of time and energy to send our people to participate in these kinds of discussions, because we want to not only have a voice, but we want to understand changes and take advantage of them," Niznik told me. "With CTOP, we became part of the process early on. Knowing that this was coming, we started developing technology on our side so that we were first in the industry to be in position to take advantage of the opportunity. We're able now to send multiple trajectory options and to more or less control our own fate. Being involved early on and making the investments, we have a deep understanding of it and we've been able to influence policy discussions, in the interests of the whole system."

If you're not convinced yet that the public needs to push for Next-Gen, take a look at the weather.

STORMY WEATHER:
THE CHALLENGE OF TURBULENCE

Flight delays and cancellations, for any reason, cost the U.S. economy over $31 billion per year, according to an FAA-commissioned study,[11] and make flying very frustrating for passengers. Delays due to weather are a big part of this. In 2013, weather caused 36.5 percent of all flight delays.[12] We can't blame those delays on extreme weather, such as hurricanes, tornadoes, or bad winter storms; severe weather accounts for only 4 percent of the delays because it prevents flying altogether. The culprit is everyday turbulence. Think of cold fronts and high winds as dynamic potholes in the sky, doing to planes what potholes in the road do to cars, or requiring costly detours. The difference is that the skies keep changing, and air routes change with them.

Delays are bad enough. Add to that the other costs of turbulence. Harm to passengers and crews from the bumps, which range from mild to severe but are harmful in any case and can lead to claims for compensation. Wear and tear on the aircraft from bouncing around. Extra fuel consumption, a big cost to the environment as well as to the airline, because pilots try to go around turbulence by making route or altitude changes that require extra fuel, adding the insult of pollution to the injury of delays. The total annual cost of turbulence to airline operations has been estimated at $13 billion, plus 5 million metric tons of carbon emissions. Associated passenger and economic spillovers include impacts to supply chains and production costs estimated at $14.7 billion, which gets factored into higher prices for products. In addition, delayed, canceled, or diverted flights raise airline operating costs in staffing and maintenance. Potential cost savings can also be quantified per flight. For example, the costs of fuel burn while taxiing amount to about $25 per *minute*; an aircraft diversion costs about $15,000 to $100,000 per *aircraft*; and an FAA tarmac delay penalty is about $27,500 per *passenger*. These numbers quickly add up to millions of dollars on a full flight.

In the unusually severe weather month of January 2014, flight delays and cancellations caused an estimated $2.5 billion in economic damage. Airlines differ in how well equipped they are to respond to these weather problems. During the two extreme storms that month, Delta and United canceled only a small fraction of their flights even though they serve storm-impacted areas, while JetBlue preemptively canceled thousands of flights. JetBlue may have wanted to eliminate any chance of repeating the incident on Valentine's Day 2007 in which a JetBlue plane was stranded on the tarmac at JFK for over ten hours. That incident was one factor in the FAA's adding that stiff financial penalty for more than three hours on the tarmac. Delta, on the other hand, had made heavy investments in technology through an in-house department, which helped predict problems and reposition aircraft and enabled Delta to avoid canceling most of its flights. Delta feels that its in-house technological edge provides an advantage, and admiring rivals at JetBlue told us they agree with Delta, as they consider their own technology upgrades.

Short of changing the weather, the best way to deal with it is to predict it accurately and to take action. The FAA estimates that two-thirds of weather-related delays are avoidable. With accurate forecasts for pilots, control towers, and operations centers, airlines could, for example, carry less contingency fuel because of their higher confidence in available routes and altitudes. With confidence in forecasts, planning could better anticipate ground holds, deicing, or capacity changes. Superior weather information would make it possible to predict airspace and route availability, delay minutes, diversions, and tarmac risk. A connected platform could alert dispatchers, crews, and stations to potential impacts, enable collaboration between dispatchers and crews to assess options and take optimizing action sooner, and reduce costs. Since weather patterns can change often, real-time data could also permit targeted midcourse adjustments that produce better results.

An American Airlines pilot said on national television, "For a long time, we have been gauging turbulence by the seat of our pants. As we unexpectedly fly through turbulence, we make an assessment based on experience." Increasingly, Big Data solutions can give pilots formidable tools besides their pants.

CLOUD COMPUTING IN THE CLOUDS

People using the term "cloud computing" don't generally mean real clouds, as in white fluffy or dark menacing patches in the sky. But it has a literal meaning for air transportation. The state of the clouds is vitally important, and the use of cloud computing services to analyze Big Data in real time for those up in the clouds offers another big opportunity for significant positive change. Enter the Weather Company as a partner.

When David Kenny became CEO in January 2012 of what was then called the Weather Channel Company, a thirty-year-old cable television broadcast company, he almost immediately sent it soaring into cyberspace. Kenny dropped "Channel" from the corporate name, signaling the simultaneous importance of its digital platforms, including weather.com. He put weather science at the center, to provide forecasts that were ever more accurate, ever more locationally precise, and ever more current, to within five minutes. He also brought the company into crowdsourcing, with the purchase of Weather Underground, a weather-for-the-people pioneer in personal weather stations, a proprietary data source beyond the publicly available National Weather Service data.

"It's a great time to be in the weather business," Weather Company executive VP and CIO Bryson Koehler said. "Weather impacts about a third of the global GDP every day. A third. Just read the first-quarter earnings statements from most major corporations, and it starts with something about weather impact on their business." He was speaking with Cisco CEO John Chambers at a Cisco meeting where Chambers praised TWC's leap to the center of the Internet of Things. Indeed, the Weather Company's key partners also include Google, Apple, and Yahoo. Weather is among the most used smartphone apps, and it's what pilots want on their iPads and tablets.

"Weather was the original Big Data problem," Koehler continued. "On March 5, 1950, the first weather model was run, using 25,000 punch cards and an ENIAC. Ever since then, weather's been going through a change, a transformation. There are 2.3 billion locations of forecast points, up from 2.2 million a few years ago, a true Big Data explosion." The Weather Company is tapping a high proportion of these data, as about 25,000 personal weather stations in the United States that feed proprietary data, compared with about 1,400 National Weather Service stations, plus another 10,000 personal weather stations that report international data affecting U.S. weather. The question is how to use the information to improve the weather forecast.

I began to see how, when I flew to the Weather Company's Atlanta headquarters, about eighteen months into David Kenny's tenure, with Kenny and Mark Gildersleeve for top management meetings. En route, on a beautiful day for flying, I heard the full story of Total Turbulence, a potentially pathbreaking solution for aviation and a boon to passengers.

Through its Weather Services International (WSI) division, which Gildersleeve heads, the Weather Company provides technology-based weather services for over 130 commercial airlines worldwide and for the FAA. Total Turbulence, one of its most promising new products, was derived from a NASA innovation. NASA had given a grant to a private venture, Aerotech, to develop turbulence detection and reporting software, called TAPS. After a period of negotiations starting in the summer of 2009, WSI licensed TAPS from Aerotech and combined it with other WSI capabilities under the Total Turbulence umbrella. Total Turbulence, which is a flexible, powerful, cutting-edge system, can automatically detect and report airborne turbulence; automatically alert impacted aircraft and pilots, including those in the trail; deliver information to the WSI Forecaster system to continually refine alerts and forecasts; use WSI Fusion, an integrated operating system, to alert dispatchers; and offer WSI Replay for postanalysis and reporting.[13]

In April 2011, the first American Airlines aircraft flew with TAPS software. By the fall of 2013, AA had deployed Total Turbulence in its 737, 757, and 767 fleets—about 350 aircraft out of roughly 700—and was starting on the 777 and new A319 and 321 fleets. "Everyone sees Total Turbulence as a game changer," AA's weather planning manager Desmond Keany says. AA found immediate savings, in the many millions of dollars and climbing, most of this from a reduction in turbulence injuries, which is a true human benefit. An AA pilot told a national TV audience that next to GPS, Total Turbulence is one of the most significant tools in the cockpit.

Other airlines are integrating WSI software into operations systems, ensuring that weather data inform decisions. For example, JetBlue wants to determine whether or not to delay or cancel flights at select airports twenty-four or more hours in advance, especially in areas where the Weather Company's winter and tropical storm expertise is essential. United Airlines wants one weather provider with global reach for on-site weather forecasting, data services, and web portals. WSI also helped United migrate to a paperless cockpit, utilizing customized tablets (super versions) to promote fuel efficiency and identify situational hazards.

Both United and American have WSI meteorologists embedded in their system operations control centers, where they occupy dedicated desks, participate in conference calls with the FAA, and ensure smooth communications and tight integration of data and decisions. One meteorologist I met in the AA center in May 2014 had worked for American for over twenty years, joined WSI when his function was outsourced, and continued to work in the same place every day, only now he had a different name on the paycheck and a link with WSI's national weather systems center.

While at American's flight operations center with Gildersleeve and Mark Miller, who manages aviation for WSI, I heard numerous testimonials about the centrality of weather and the importance of coping with turbulence. Better decisions in regard to turbulence make a difference for passengers. "From a passenger point of view, we've only had one injury incident in the last four months. That's very good, industry-leading—knock on wood," Desmond Keany said, knocking on the table. "I don't think that's any one technology, but that's everything together, from iPads to how we operate to the rules for flight attendants."

"Injury-free" is among the nonevents that don't make headlines. "I arrived, I didn't get hurt" sounds barely a notch over negative. It's hard for people to feel joyous about a nonevent. Relief maybe, but not enthusiasm.

I told the AA group about my turbulent AA flight from Boston to

Dallas–Fort Worth that day and wondered how Total Turbulence entered in, since we had a long delay. Ray Howland got on his laptop and projected a map on a big screen, where we could see all flights moving in real time but could also go back to find the path of my flight. We looked at the weather swirling around the flight path: a cold front in the Tennessee Valley heading east. I told them what was happening on board. Takeoff in Boston was delayed thirty minutes to load extra fuel for rerouting around turbulence, and service was delayed for forty-five minutes after takeoff. The captain came out of the cockpit and told us, in a candid and thorough way, about the delay and the reasons. Yet the flight was still a bit bumpy. That it could have been much worse didn't stop people from being disgruntled by the late arrival, especially the passengers pushing through to make a tight connection.

Brian Will was listening and put a different spin on this kind of story by offering a pilot's perspective. "I was flying a 777, and we had to take a delay because of the weather," he said. "I got on the big voice on the airplane and said, 'Ladies and gentlemen, we've got some weather here, so we're going to take a delay.' You could hear groans from the back of the airplane. So I waited a second and said, 'I heard that.' Then I said, 'You would much rather be here on the ground wishing you were up in the air than to have me get airborne and suddenly find yourself up in the air wishing you were back on the ground. Trust me, when it's safe, we go. If it's not, we don't. Over.' I hung up. The flight attendant came to tell me, 'Everyone liked what you just said. They've changed their minds all of a sudden.'"

Information empowers. Empowered pilots in empowered airlines can empower passengers, or at least enlighten them. Real-time decision-making can reduce costs and minimize delays. That accounts for the enthusiasm for getting NextGen technologies ASAP. So what happens if major parts of NextGen are delayed or scrubbed? Advocates say there is no Plan B.

NextGen also supports surface collaboration of aircrafts and other ground support to optimize efficiency at the airports. One successful example was the use of NextGen technology during the expansion of the longest runway at New York's John F. Kennedy International Airport (JFK), known as 31L/13R. According to the FAA, to minimize disruption to JFK operations during 31L/13R's expansion, the airport used NextGen's departure queue metering system, which is a technology through which departing aircrafts would be assigned a takeoff time and location and wait at the gate until takeoff. The project significantly reduced taxi congestion around the runways and continued in use after the 31L/13R expansion was completed. From this experience, the FAA estimates that nationwide use of the technology could potentially save 5 million gallons of fuel and 7,000 hours of taxi time per year.[14]

But while the air side is working on NextGen, the land side beyond flight operations often seems stuck in LastGen. Airports pose a problem for the United States that is ripe for rethinking.

AIRPORTS, THE GROUND RULES

The goal for airlines is to get people out of town fast. The goal for airports is to keep them on the ground spending. Cities and towns see dollar signs as they seek the economic benefits of airport development that can keep money, if not the people, in their region. Airports are sometimes more about the land than the landing. A clustering of activities around an airport, whether hotels for passengers or warehouses and distribution centers for cargo, can create jobs and perhaps even a new wave of future settlement dubbed an "aerotropolis," as a catchphrase for a metropolis centered on an airport.[15]

U.S. airports provide service to more than 800 million passengers annually and will probably reach the billion mark annually by 2027 or sooner.[16] The hub-and-spoke system utilized by many airlines has concentrated passenger air traffic in several main hubs; just fifteen airports handle 49 percent of all passenger boardings. In the United

States, delays at these hub airports cause a ripple effect through the country's aviation system.

There are 503 commercial airports in the United States, of which 382 are primary airports providing passenger service. Of the primary airports, 29 are classified as large hubs, 37 as medium hubs, 72 as small hubs, and 244 as nonhub airports.[17] U.S. hub airports, compared with their counterparts abroad, consistently receive lower marks for customer service, feature more delays and congestion, and have older infrastructure. There are many big airports in the United States, but few are considered good enough by world standards.

On the 2013 version of the Skytrax annual survey of the world's best airports, 12 million airline passengers ranked over 400 airports around the world across 39 different categories. No U.S. airports were in the top 10, the top 20, or the top 25. Only 4 U.S. airports squeezed into the top 50, with another 11 in the top 100. That's out of 382. The highest-ranked U.S. airport on this survey, taking all factors into account, is Cincinnati/Northern Kentucky, clocking in at number 30. Denver ranked number 36, San Francisco number 40, and Atlanta Hartsfield-Jackson number 48. The others in the top 100 are Dallas–Fort Worth, Seattle-Tacoma, New York's JFK, Minneapolis–St. Paul, Detroit, Chicago's O'Hare, Raleigh-Durham, Charlotte-Douglas, Boston's Logan, Salt Lake City, and Pittsburgh.[18] The striking thing here isn't just the airports (Cincinnati doesn't spring to mind first as a major global destination), but that 29 airports abroad are considered better than any single U.S. airport. Of course, that's not a surprise to those of us who fly internationally.

Nations in Asia and the Middle East are developing new, efficient, technology-enhanced airports that look like high-end shopping malls full of leisure amenities, but American airports are falling behind in core infrastructure areas as well as in amenities. According to the Airports Council International–North America, airports require $14 billion per year to fund existing capital projects, and they must find more than $80 billion for new infrastructure to meet projected growth in air passenger traffic through the early 2020s.[19] That doesn't include upgrades in airport communications to ensure that pilots, dispatchers, and air traffic controllers can always be on connected networks to share data.

Getting to and from airports is another pain point in the United States. In many larger cities, airports are often located fairly far from the city center. These distances contribute to traffic congestion on the ground from autos, vans, and buses; rail service to airports is still a rarity. Numerous weary travelers join experts in wanting more and better intermodal connections to ultimate destinations from the airport, including high-speed rail from inside airport terminals rather than remote locations. They've seen it in Tokyo, where passengers can board a bullet train in the airport. Or at London's Heathrow, where frequent fast trains to London's Paddington Station are just a floor below the arrivals area, and so is the Underground, the subway to the city. Getting even more utopian, some Americans dream of a single ground-to-air-to-ground ticket, maybe through a bar code on a smartphone.

WHO DECIDES AND WHO PAYS?

Decisions about airports are highly decentralized in the United States. Although states have the constitutional power to make such decisions, state constitutions typically leave decision-making to local communities, through local or regional airport authorities. They can also be highly politicized. Airport projects are not just a business matter; they affect numerous stakeholders in surrounding communities, with concerns about noise, carbon emissions, open spaces, impact on the natural environment, or acquisition of off-site property. Airlines themselves do not always favor airport expansion plans, especially because airlines will bear the costs in ticket fees.

Airlines pay by passing fees on to us, the passengers. To fund airport operations and improvements, cities or regional airport authorities rely on airline fees, passenger charges, the sale of municipal bonds related to passenger traffic, and some federal taxes. The federal Airport and Airway Trust Fund, created in 1971, enables the FAA to support air traffic control and some other airport services, drawing from a 19.4 cents per gallon aviation gasoline tax, a 7.5 percent ad valorem tax on domestic commercial ticket prices, a $3.90 per segment flight fee, a $17.20 per passenger international arrival and departure tax, a frequent-flier tax, and a 6.25 percent tax on payments for air cargo. Airline ticket fees account for 70 percent of the revenue collected, including a passenger facility charge authorized in 1990. Although they generate billions of dollars a year, such taxes have historically funded only 30 percent of airport capital improvement projects and would fall far short of the need; but an increase in federal taxes is not on the agenda, since they already account for approximately 20 percent of the average airline ticket price, 7 percent greater than 1992 levels and 10 percent higher than taxes on "sin products," such as firearms.[20] Legislators are considering doubling the 9/11 security tax as a means to increase revenue to fill general budget shortfalls, not for an aviation expense. "Airlines are the most taxed industry in the country," complains Delta CEO Richard Anderson.

Moreover, in many cases airlines pay for terminals that carry airline brands. JetBlue's vice president for airports, Ian Deason, says that business is won or lost on the ground. JetBlue's investment in Terminal 5 at JFK Airport in New York is a case in point: a gem in an otherwise run-down airport. American Airlines made a large investment in Terminal D at Miami International Airport (MIA), although the county had to step in to complete it. Terminal D is actually A–E, a joining of former Terminals A, B, and C to Terminal D, with luxury shopping and overhead light rail connections between gates, plus a walkway to the older, not yet refurbished Terminal E, which is still called Terminal E. But there is no Terminal A, B, or C. Confused yet? So are many passengers trying to find their way around MIA.

Take away the airline contribution, and watch airports seek new sources of revenue by using the real estate differently. Pittsburgh's airport was among the first in the United States to increase retail space and attempt to become a destination shopping mall for local shoppers, but post-9/11 security restrictions have dampened similar plans. High-end retail space at most airports serves only passengers behind the security walls. There are efforts to turn areas around airports into warehouse centers, much as Memphis did after Federal Express became a cargo success story. Denver International Airport has installed solar panels on unused land to reduce its electricity expense. The Dallas–Fort Worth airport is seeking to boost nonaeronautical revenues, including retail rentals and oil and gas production on airport land. The airport has 6,000 acres of undeveloped land that might become business parks. Airports are big business and real estate plays. Sometimes it seems that flying is a sideline.

THE POLITICS OF AIRPORTS

It can take five, eight, or more years to build a new terminal and even more to complete a runway, for political, not technical, reasons. In Chicago, modernization at O'Hare Airport, one of the nation's busiest, was championed by the business community starting in 1999 and yet took years to win political support from all levels of government before it took off in 2008. Seattle-Tacoma Airport took thirty years to add a third runway because of local opposition, largely for NIMBY (not in my backyard) and environmental reasons. Delta Air Lines' Anderson observes that "the environmental process is pretty tough on runways." Review criteria include noise, air quality, water quality, visual impacts, fish, wildlife, plants, floodplains, and wetlands, not to mention the numerous concerns nearby residents can imagine.

O'Hare (ORD, in its aviation code) is among the top twenty busiest airports in the world with about 1.5 million tons of imports and exports, worth approximately $115 billion each year. O'Hare is in close proximity to the rail yards and interstate highways that are located within a one-day truck drive of 29 percent of North American consumers and within a two-day drive of 42 percent of consumers.

ORD was originally developed in the 1950s, on the periphery of the city and spanning two counties, to relieve growing congestion at Chicago's in-city Midway Airport and to help increase mobility in the region. However, as air traffic expanded in the 1970s and

1980s, O'Hare did not. By the mid-1990s, the airport was operating at capacity. Any disruption in service, especially weather related, strangled air traffic in Chicago and sent a wave of delays across the national and international aviation system.[21]

In 1999, a Chicago business group, the Commercial Club, commissioned a study showing the need to modernize ORD. As an economic engine for Chicago, ORD was said to contribute $37 billion annually and over 400,000 jobs and could potentially create an additional 195,000 jobs and $18 billion in annual economic activity. Airlines could save $375 million each year, and passengers could collectively save $380 million per year. Yet even with all these billions on the table, Mayor Richard Daley initially declined support. In 2001, on a trip to Chicago, Senator Tom Harkin of Iowa heard about the airport's growing congestion and capacity constraints and proclaimed that the federal government would respond if the City of Chicago did not take action. The Commercial Club bought two pages of advertisement in all Chicago newspapers; more than fifty companies pledged their support, calling for additional local, state, and federal backing.

Mayor Daley eventually embraced the project. But it still needed support from various constituencies: Republican Governor George Ryan; DuPage County's leaders, who had long opposed any airport expansion; other surrounding communities; and major airlines operating at ORD. Suburban residents didn't see what was in it for their communities. More politicking ensued. After local support was secured, airport modernization required FAA approval. In 2003, the FAA opened a regional office in Chicago, expediting regulatory processes.

Although airlines are responsible for the funding of airport infrastructure, as I've indicated, they don't always find enough demand to justify runway expansion and they question public officials' and local business leaders' motivation. "Politicians want to overbuild and make something that looks impressive and expensive, since the airport is the gateway to the city," one airline executive in the region told us anonymously. Early negotiations between Mayor Daley and the CEOs of the major airlines at ORD, United and American, stalled, and the airlines sued to delay the work. U.S. Transportation Secretary Ray LaHood stepped in and brokered a deal in 2011; the airlines would drop the lawsuit, and the city agreed that only one more runway would be built, not the whole project plan.[22] Financing of nearly $1 billion came from airlines through passenger facility charges, as well as general airport revenue bonds and Federal Airport Improvement Program funds, not from local or state taxpayers. The higher fees concerned the airlines, which were struggling to recover from the Great Recession. But business associations pointed to benefits to Chicago from attracting new North American corporate headquarters —Siemens, GE Transportation—because of the airport.

When Mayor Rahm Emanuel took office in 2011, he announced that ORD would receive a $1.4 billion investment over the next three years to open two new runways to be completed in 2013 and 2015, also from passenger fees, grant programs, and municipal bonds, rather than taxpayer dollars. This would create 5,900 jobs and secure direct connections to Asia, Latin America, and Europe, reduce delays by 80 percent, and raise the capacity of the airport by 300,000 passengers (60 percent) each year by 2015.[23]

Emanuel's announcement received mixed reviews. Airlines agreed that the new runway completed in 2013 along with the runways from four years earlier significantly helped resolve delays. However, airlines questioned the business need for an additional runway planned for 2015. Observers also wondered why targets for modernization did not include greater use of technology for passenger and cargo enhancement, better intermodal connections such as train stops in airport terminals, and reduced traffic congestion on the ground en route to O'Hare. Emanuel argued for the immediate pain relief and the future opportunities, and business interests agreed.

REGIONAL COALITIONS AND CONFLICTS: THE LAND BEYOND AN AIRPORT

If you die and go to heaven, the joke says, you must connect in Atlanta. Hartsfield-Jackson International Airport (ATL), home hub for

Delta Air Lines, is the world's busiest airport, serving about 90 million passengers a year (although Dubai is hard on its heels). Owned and operated by the City of Atlanta, it is bordered by other small cities and two counties. It provides 58,000 jobs on the airport grounds, supports 403,000 jobs regionally, and claims a $32.6 billion economic impact on the region.

In 1999, airport leaders developed a $6 billion capital improvement plan. A master plan advisory committee provided input and helped secure community buy-in for a wide range of projects now in various stages of completion, including reconstruction of inbound roadways and a new international terminal. The Metropolitan Atlanta Rapid Transit Authority (MARTA) offers direct bus and train service to the airport, a seventeen-minute ride from MARTA's rail station hub. That rail service makes Greater Atlanta a standout among U.S. cities.

Civic coalitions combining local governments with business and nonprofit leaders have sprung up to maximize economic development around the airport. In 2011, the Atlanta Regional Commission; in 2012, the Airport Area Task Force; in 2014, the Atlanta Aerotropolis Alliance. The German auto manufacturer Porsche was lured to invest in a parcel adjacent to the airport for its North American headquarters. Who needs the air, when you have airport land?

ATL's congestion and delays are notorious; an ambitious master plan for 2030 faces capacity constraints. Atlanta is the nation's ninth-largest metropolitan area but the only city in the top ten that lacks a secondary commercial airport.[24] Why not a second airport instead of one megaproject? Perhaps interests in minimizing competition explain why not. In 2013, a single runway airport with one gate, located thirty miles northwest of downtown Atlanta, was in the process of obtaining government approval for commercial airline service. Delta opposed those plans; an executive wrote that "a second airport can quickly expand, and the impact on Hartsfield-Jackson would be significant." Hartsfield's general manager told the press that a second airport "isn't commercially viable or economically sound for the region." Atlanta mayor Kasim Reed added that "the overwhelming amount of data indicate that having two airports is unhealthy."

Whether or not more than one airport is unhealthy, the competition around secondary airports could be. In Greater Los Angeles, there are complaints that congested Los Angeles International (LAX) has, regardless, interfered with airline service at uncrowded, underutilized airports nearby. But LAX champions argue that there is insufficient demand for those airports, which sprang up without a regional strategic vision. That's true too. American airports derive from local, not national, hopes.

Karen Freeman-Wilson, mayor of Gary, Indiana, sees a run-down regional airport as a source of jobs and economic development for her Rust Belt community. The airport, near the Illinois–Indiana border, is under thirty miles from downtown Chicago, a short hop via expressways. There's even Chicago in the name: Gary-Chicago International Airport. Why not promote it for passengers on the South Side of Chicago and address congestion and capacity concerns at ORD? One problem: it has been an airport without an airline. In mid-2013, Allegiant Air, the last of six airlines to have come and gone from Gary since 1999, flew its last flight from Gary, citing weak demand. In January 2014, approval was secured for a forty-year public-private partnership with AvPorts, a subsidiary of Aviation Facilities Company in Dulles, Virginia, for expansion of the airport, and a runway that could accommodate bigger aircraft.[25] Freeman-Wilson championed the project and worked with numerous Chicago officials on collaboration. But church groups in Gary protested a lack of provision for jobs for the less-advantaged. And a Chicago official told me off the record that chances were small of diverting traffic from Chicago. But Mayor Freeman-Wilson could yet be proved right; leaders press on despite skeptics.

THE NONSTARTER: PRIVATIZATION

One financial ploy that's not taking off in the United States is airport privatization. Oh, there are dreams: a deep-pocketed investor takes over, provides cash to balance city budgets and long-term capital for infrastructure improvements, operating without bureaucratic constraints. But those are pipe dreams. Although cities are willing to sell airports, no one wants to buy them. Surrounding real estate might be valuable, but the terminals and facilities don't seem to be.

The tiny Stewart International Airport in Newburgh, New York, was the first commercial airport to participate in an FAA pilot program to encourage privatization. It didn't last. Stewart was privately held from 2000 to 2007, when the Port Authority of New York and New Jersey bought it back from the UK's National Express Group, which had decided to exit the airport business.[26] Next, in February 2013, Highstar Capital, an infrastructure investment firm, together with the Mexican group ASUR, took San Juan, Puerto Rico's Luis Muñoz Marín International Airport private in a forty-year, $615 million concession, giving Aerostar Airport Holdings a forty-year operating lease for the Caribbean's busiest airport. A newly elected governor promised not to hinder the already completed deal despite his personal opposition— a common political risk for infrastructure deals is a change of administrations and loss of official support.[27]

This is hardly a gold rush. Another pending application (of a mere two) is an economic development deal for Florida's Hendry County, one of the state's poorest, to turn Airglades Airport into a cargo hub that could divert cargo from Miami International, freeing up space for more lucrative passenger traffic. The potential buyers are a local consortium in the agriculture business: U.S. Sugar, Hilliard Brothers, and Florida Fresh Produce. And that's about it. Seven other airports, including New Orleans' and San Diego's, applied to the FAA program but withdrew. Chicago tried to privatize Midway Airport twice but failed as a result of limited investor interest. In the latest attempt, under Mayor Emanuel, there were only two bidders. Then one, unable to raise the needed capital, dropped out, and the offering was canceled.[28]

One alternative is public-private partnerships for individual terminals rather than for the whole airport. This is being explored by the Port Authority of New York and New Jersey as it considers bidders for a thirty-five-year concession to design, build, operate, and finance the central terminal at New York's LaGuardia Airport. "This could be interesting," says former FAA Administrator Jane Garvey. "If it works, Newark or Denver could be next. The airports are watching New York's experience."

Meanwhile, the public sector is stuck with airports no one wants to buy, financed by airlines that feel overtaxed, and populated by passengers who would rather not spend time in them at all. And this is all local and regional. The hunt for capital and the lag in airport infrastructure modernization signal the need for new national strategies in the United States. This is essential because air transportation connects us not just within our homeland but across the world. On the international dimension, the United States is falling short and falling behind.

THE SKIES OVER THE SEAS

International standards and elaborate agreements have long characterized the sector, even while national flag carriers dominated their home markets and travel in and out of their countries, and airspace was closely protected by each country. Mergers have generally stopped at national borders, owing to regulations restricting foreign ownership. As an alternative, airlines have developed international alliances with foreign-based carriers. American Airlines developed the first, the One World Alliance with British Airways, Qantas, and others, to ensure access to a broad range of destinations by means of code-sharing agreements, along with opportunities for interchangeable mileage awards, coordinated baggage handling, and joint premium services.

Open Skies agreements among governments and free-trade pacts allow international carriers access to U.S. destinations, the largest air travel market in the world. The rationale is that it also opens markets for U.S. goods and opportunities for U.S. citizens. Such agreements date back to a framework set in 1944. The United States began pursuing Open Skies agreements in 1979, but this really took off with agreements in Europe in the 1990s. There are one hundred bilateral agreements, the State Department indicates, and some multilateral agreements with several countries at a time: in 2001, the United States signed a multilateral agreement on the liberalization of international air transportation with New Zealand, Singapore, Brunei, and some small Pacific nations, and in 2007 with the European Community and its twenty-seven member states. But many U.S. airline executives feel that Open Skies frameworks haven't caught up to recent developments.

Some international competitors have benefited from national strategies of investment to make their airports hubs for international traffic—a successful strategy for Singapore in that it made Singapore Airlines among the most highly ranked passenger airlines in the world and drew connecting passengers through Singapore's Changi Airport, helping develop a new nation. Singapore Airlines, flagship of a country with a tiny 5.4 million population, is among the world's largest airlines in market capitalization, which is a tribute to the international strategy. Like many entities important to Singapore's development, the airline is "government-connected," majority owned by Temasek, the Singapore government's holding company.

Taking off from the Singapore model, Middle Eastern countries have developed similar national airline and airport strategies. For the United Arab Emirates (UAE), subsidizing airlines and airports is part of a strategy to diversify away from oil, seeking to create international hubs attracting high volumes of passenger flights through airports, connecting traffic to and from Australia and Asia through Dubai and on to Europe and North America, while hoping that a portion of the people and their money will stay in Dubai. Carriers such as Emirates (Dubai), Etihad (Abu Dhabi), and Qatar Airways enjoy the benefit of an estimated $10 billion to $20 billion in airport infrastructure (financials are not released), with dedicated terminals, low airport charges, the most technologically advanced services, an absence of labor unions, no corporate income tax or fuel tax, and a major emphasis on duty-free retail shopping.

A knowledgeable airline executive who spoke off the record said that the cost to land in the UAE is one-quarter of what it is in the United States. Moreover, Middle Eastern airlines have made large investments in fleet growth, with 355 wide-body jets ordered by Emirates, Etihad, and Qatar, at an investment the expert estimates at $160 billion, complete with the latest in-flight Wi-Fi and luxury, adding up to more high-tech aircraft than all U.S. carriers combined. GE CEO Jeffrey Immelt, whose company is a supplier, comments, "When I was a young GE guy and I wanted to fly anywhere around the world, I always flew to London. Nobody ever flies to London anymore because of Dubai. The focus by the Emirates government to become the centerpiece of the world, and their willingness to invest in infrastructure ahead of anybody else, changed the game." Representatives of U.S. carriers agree, and they suggest that international Open Skies agreements are long overdue for updating. They stop short of crying "unfair competition."

That's not all that suggests a rethinking of American national strategy toward international travel. Personally, I thank U.S. Customs and Border Patrol every time I go from plane to curbside in a few minutes thanks to Global Entry kiosks, where a scan of my passport and fingerprints speeds me and my carry-on through immigration formalities. Not so for foreigners. Cumbersome U.S. policies and processes make travel difficult, if not impossible, for many potential international visitors. At 11 percent of total airline passengers, international travelers contribute over $116 billion in direct spending and $1 billion in indirect spending annually in the United States. But the World Economic Forum ranks the United States at number 121 of 180 countries in terms of the burden of visa requirements. More than 75 percent of the potential travelers from China, India, and Brazil responding to a survey believe that it is difficult to impossible to travel to the USA; approximately one-third are deterred by the visa process. Those who are able to travel to the United States encounter long wait times for non-U.S. passport holders in immigration and customs lines, which peaked in 2012 at over four hours at New York's JFK Airport, and nearly four hours at Miami, O'Hare, and Dallas–Fort Worth—all after a lengthy international flight.[29]

On the cargo side, international competition is starting to challenge U.S. dominance in global shipments. Hong Kong has already replaced Memphis as the world's number 1 air cargo hub. In 2013, Emirates SkyCargo, based in Dubai, overtook Federal Express for the number 1 position in freight carried, according to informed inside sources in advance of data release.

If you can't beat 'em, join 'em. One way to deal with international competition, and international opportunities, while learning what is state of the art in other places, is for U.S. airports to woo international airlines to offer nonstop service and easy connections. Massport, the Massachusetts Port Authority that runs Boston's Logan Airport, is doing this very effectively. Emirates and Etihad began offering nonstop service from Logan to Dubai and Abu Dhabi in 2013, and in the late spring of 2014, Turkish Airlines and China's Hainan Airlines started wildly successful flights from and to Istanbul and Beijing, respectively. Cathay Pacific to Hong Kong is coming.

I flew to Istanbul during the second week of the new Turkish Airlines nonstop Boston–Istanbul–Boston service for a short business trip that wouldn't have been possible without the nonstop flights. While there, I met with Turkish Airlines CEO Temel Kotil, who had joined in 2005, just after the airline was privatized by the Turkish government. That started a rapid ascent for Turkish from mediocre or worse to the very top. Now number 1 in the world on indicators such as service and flying to 257 destinations, more than any other airline by country, Turkish represents a noteworthy success story. But Istanbul's international airport remains less developed, to put it politely. Turkish plans to build its own terminal, with the latest features, so that it can compete with Emirates on the ground as well as in the air.

Turkey has a gift-exchange culture for token mementos. Mr.

Temel gave me, at the end of our meeting, a large model of the Airbus A330 that flew me to Istanbul. It was too bulky for my carry-on luggage, so he sent it to me later along with Turkish Delights candy. They were hand-delivered by the airline's Boston general manager, Adnan Aykac, who stayed in my office to talk about international comparisons while my staff munched the candy. Boston was already looking like a winning destination for Turkish; flights in the first month (May–June 2014) were 90 percent full. Aykac praised Logan's efficiency; because of limited space, Logan required fast turnarounds of aircraft and knew how to make that work. Turkish also valued code sharing with JetBlue, for seamless ticketing internationally and domestically was important, and JetBlue had the most flights out of Logan to U.S. destinations. But in addition to Logan, Turkish flies directly to other U.S. destinations, and Aykac identified many areas of backwardness compared with his experience at numerous international airports.

"The U.S. needs to improve," he said, trying hard to be diplomatic. "Being big isn't enough. You have to plan airports to operate efficiently. You need continued technology investments to modernize systems. Air traffic control is sufficient but could still be improved. Security issues can cause two-hour delays. . . . The U.S. needs to improve," he repeated. Numerous American passengers echo that sentiment.

TAKING OFF TOWARD A BETTER FUTURE

American private-sector innovations have made air transportation possible and have the potential to modernize and improve every aspect, while government policies have shaped it every mile of the way, through investments in defense, creation of a central federal regulator, and the actions of local/regional airport authorities. But the interactions of those players, and the tensions between innovation and bureaucracy, have not always been handled in an optimal way. Throughout this chapter, the same themes fly by: creative innovators, including pilots like Brian Will, on the forefront of new developments; bureaucracies that move too slowly; regional politics often substituting for national strategies. Technology is speeding ahead; regulators struggle to catch up. The land side is governed locally; the air side is governed nationally. It is time for leadership at the national level to create a national strategy for making air transportation better for all who use it. The industry association Airlines for America has been calling for it.

In a domain for which technology has always been critical, the development of NextGen could produce significant benefits despite the controversies, reduce negative impacts, and maybe even do something about the weather. That doesn't mean uncritically embracing every proposed technology bell and whistle, but it does mean testing, deploying, and modifying faster, coming closer to the rapid prototyping of the technology world, which permits in-use adjustments, while regulators join industry teams to ensure safety and cost-effectiveness.

Within the United States, delays and other ills of a less-than-optimized system have become routine annoyances. But it doesn't have to be that way. Imagine flying with pilots empowered by technology to make better decisions for passengers. Imagine NextGen air traffic control generating the quintuple wins in greater safety, lower costs, fewer delays, less carbon, and seamless connections between many places for future opportunity—all with noise-free landings. This could free resources to be invested in improving infrastructure. It could help create public appetite for modernization. Technology in the air and on the ground is not just nice to have; it's a must-have. It could restore American preeminence in air transportation and improve economies and lives.

The FAA should move faster to consider long-term investments and speed up approvals. The idea of an independent entity for air traffic control should be revived and freed from the short-term congressional budget cycle—which would generate tension even if Congress were not dysfunctional. The FAA and other agencies within and outside of the U.S. Department of Transportation should collaborate on moving promising technology forward faster, especially communications technology, to accelerate NextGen.

The problems are not just in the air; they are connected to the ground. Being able to fly with fewer delays won't be enough if there are also major delays getting to airports and inefficiencies at airports themselves, including outmoded communications in shabby facilities using decades-old processes. An examination of U.S. airports shows them to be more LastGen, in contrast, mired in regional politics and failing to meet international standards even as international competition grows. Because of the way airports are financed, passengers ultimately pay the price. It is important to ask why private companies have not wanted to operate U.S. airports. Overall, decisions are made locally without regional considerations and national strategies. Improvements are coming gradually, as intermodal connections are envisioned, and the FAA works more closely with major airports as well as with telecom for communications and with public transit—rail and buses—for ground connections. But the pace is much too slow. If leaders and the public can get the 10,000-foot view of the system as a whole, progress can be accelerated.

American institutions too often operate in silos. Air flies above all of them. Building a national strategy with diverse independent stakeholders is difficult, but if the public demands it, it can be done.

4

SMART ROADS MEET
THE SMARTPHONE

Transportation by road must get smarter. Developments in the past decade—smartphones, sophisticated sensors, GPS, cloud computing, and Big Data analytics—are challenging old business models and causing industries to collide. Roads are coming to life. Vehicles don't need drivers. Users don't have to own vehicles. And wireless networks can manage the whole system.

These developments in technology are aligned with an equally rapid emergence of a new economy of sharing, not owning, and of services overshadowing products. It's an ecosystem that entrepreneurs target for high-growth businesses, but one that requires public-sector partnerships and public debate about how to govern it.

One often mentioned hope for technology is that it will make us *less* mobile, taking traffic off the roads. We'll stay home and work. We'll take service calls or hold meetings via Skype, Cisco telepresence, or VGO robots, and we'll get refills of disposable diapers from 3-D printers. In fact, a recent study showed that users of broadband services, mobile or fixed-line, drove an average of 67 fewer miles per month (about 800 miles annually) thanks to conducting work online.[1] But societal patterns are not determined by technology alone. Bosses still want to monitor employees, co-workers still want to look each other in the eye for team planning, and offices are useful places for serendipitous encounters—the reason why, early in her tenure, Yahoo CEO Marissa Mayer banned remote work. Ponder that. A tech company insisting on face-to-face.

At the intersection of digital technology and transportation, change is driven by a sweeping range of players across industries and sectors: telecom providers, auto manufacturers, transportation operators, IT companies, shippers, energy providers, Silicon Valley–type entrepreneurs, sharing economy activists, and national, state, and local government agencies. Unless we're also smart about it, it will be one big traffic jam.

Travel on the Information Superhighway doesn't repair deteriorating physical roads, restore crumbling bridges, or make potholes disappear, but it can speed up the process. Roads have supposedly been the main focus of American surface transportation policy since the 1950s, with the construction of the Interstate Highway System and accompanying gasoline taxes, but they have not been hot spots for innovation, and, for the most part, neither have the industries riding on them, which became entrenched and slow to change—from big auto companies to small taxi companies. In the 2010s, consumers' new love affair is with their smartphones, not their cars. In every aspect of the ecosystem—roads, vehicles, users, and information networks—wireless technology offers welcome possibilities to alleviate the pain points and bottlenecks on America's roads. The possibilities for renewal and reinvention from significant technology shifts could arouse a public appetite for repairs, if every fix-up project adds technology.

This arena is a work in progress. It's hard to change one element without affecting the others. They either connect or collide. But as a starting point, each can get smarter.

ROADS: COMING TO LIFE

The oldest class of infrastructure—the Romans built roads—is hardly known as a platform for innovation. Complex interstate highway systems require advanced civil engineering talent, and surfacing materials might be the best and latest, but they're still just inert surfaces, in the same family as dirt paths. New construction techniques such as "instant bridges" are intriguing because they save time and involve minimal disturbance to the areas where they're built—they can be constructed mostly off-site and rolled into place in a few days as permanent fixtures intended to last for decades. I saw them in Chicago and around Boston. They're built more efficiently with the latest materials, but they're still just bridges.

Add a digital dimension, however, and roads spring into life. Highways are increasingly dotted with sensors and capable of changing on the basis of data analysis. Once inert physical surfaces are becoming intelligent, flexible, dynamic platforms that can deal with problems such as congestion and traffic management, pricing and toll collections, safety and maintenance. Roads can even arrange to get their potholes repaired. Street Bump, a smartphone app created by the City of Boston, identifies potholes from bumps of the car as users drive, and immediately transmits the location to the city. Imagine that: your car and your phone are allies in road repairs. The Information Superhighway reinvents the highway.

Road enhancements date to the industrial age. To manage carriage and wagon traffic, the world's first traffic light was installed in 1868 outside of the British Parliament. It was a manually operated, gas-lit red-green system for use at night; it lasted less than a month before it exploded. By the turn of the century, manual two-color lights were in use in a few U.S. cities but were small and unsophisticated. It was not until 1920 that William Potts, a Detroit police officer, introduced the modern three-color electric light at a four-way intersection but didn't patent it. Garrett Morgan, an African American inventor and entrepreneur in Cleveland who is also credited with the gas mask, invented a slightly improved system and received a patent in 1923. He later sold rights for his design to General Electric, and it became the foundation for subsequent development. In his honor, the U.S. DOT under Rodney Slater created the Garrett A.

Morgan Technology and Transportation Futures program for education and career skills.

Potts's invention, however unheralded, was one of many firsts for Detroit and the region as the capital of the emerging auto industry. Michigan built America's first mile of concrete highway in Detroit in 1909, created the nation's first painted centerline (1911), roadside park (1919), practical highway snowplow (1922), and superhighway (1923), an eight-lane divided highway. Not surprisingly, Michigan was the first state to complete a border-to-border interstate highway under the 1956 Interstate Highway Act. You could say that this paved the way for auto hegemony.

In 1960, a radio-controlled traffic light was patented that could be used to adjust the signal to traffic. In 1967 an advanced computer system was installed in Toronto with better traffic detection.[2] Computers began to be an important factor at the same time that GPS satellite-based systems were developed. But it took until the advent of widely used wireless networks in the 1990s to make surfaces come alive with information that helps Americans use roads more effectively.

Pick your pain point. In northern California, that's likely to be the freeway maze at the east end of the San Francisco Bay Bridge, where two interstate highways, I-80 and I-550, converge, holding six to eight lanes of potential bottlenecks. This is one of the most highly trafficked areas in the region and beyond, carrying 268,000 cars daily to San Francisco and the East Bay. Caltrans, the California DOT, needed a better vehicle detection system than the then current video/radar system involving wire loops. And if Caltrans stuck with the traditional system, it would require digging trenches across multiple lanes, even on the bridge; minimizing lane closures was another requirement. Wireless to the rescue! Sensys Networks, founded in 2003 in Berkeley, California, embedded low-maintenance wireless sensors in the roads and under the bridge with minimal disruption and a ten-year battery life. (Hmm, can we get those for our smartphones?) A total of 561 vehicle detection system sensors later, Sensys won the 2008 best-innovation award from the Intelligent Transportation Society of America. Similarly, in Baltimore, the city embedded a thousand all-weather Sensys sensors without having to hire contractors, thus saving taxpayer money, and with only twenty minutes of lane closure each, thus saving driver and passenger aggravation.

We can no longer say (paraphrasing Gertrude Stein) that "a road is a road is a road." Thanks to sensors, the same stretch of fixed, passive roadway can be a different entity depending on the time of day, the amount of traffic, or the weather. The Finnish company Vaisala's weather sensors built into roadbeds can identify maintenance needs as well as snow and ice, heavy rain, fog, high winds, and sandstorms (the last undoubtedly desirable to their African and Middle Eastern customers).

Above-ground sensors from Econolite Control Products integrate cameras and processors. Sensors tell drivers how fast they're going, or tell traffic lights when to turn red or green. They can keep vehicles in lanes or help cars find optimal safe speeds or provide slipperiness indicators.

PAY AS YOU GO, FASTER

High-occupancy toll (HOT) lanes are another increasingly common technology-empowered driving option to reduce congestion, which also generates highway revenues directly from drivers. Want to go faster with less traffic? Then pay more.

Economists have long argued that user fees are an efficient way to finance transportation operations, especially when fees are higher for longer distances or travel during peak hours. This is called "dynamic pricing," pricing that changes depending on time of day and use. It was formerly called "congestion pricing" until politicians realized that the phrase combined two words the public detests: "congestion" and "pricing." (This observation comes from former New York Deputy Mayor Stephen Goldsmith.)

HOT lanes are like auctions for the use of road space, brought to you by sensor-generated Big Data. You can play this game for road segments in and around Seattle, Miami, Houston, Los Angeles, Austin, Denver, Washington, DC, San Francisco, Minneapolis, and many others, saving you time if you're willing to pay enough—or, if you don't want to pay, encouraging you to travel at less-congested hours. A study in Washington State showed that use is spread pretty evenly across income brackets, allaying concerns that only the affluent can afford HOT lanes, while everyone else is late to work.

Minneapolis–St. Paul pioneered the use of network technology and data analytics to manage highway traffic. To bypass traffic congestion, motorists in the region are willing to pay a whopping premium—the hourly equivalent of $60–$124—for use of an HOT lane. The system, informed by sensors placed along the highway, dynamically changes pricing on segments of I-394 and I-35W from twenty-five cents to $5, depending on time of day and use. Rates appear on electronic signs. Charges go to MnPASS transponders inside vehicles, to be deducted from prepaid accounts as the highway sensors detect users.[3] Other highways with even more congestion, such as the infamous Beltway around the nation's capital, have moved quickly to implement their own systems; in 2013, the Intelligent Transportation Society of America gave the 495 Express Lanes, a partnership of the Virginia DOT, Federal Highway Administration, and private operator Transurban, a "Best of ITS" award.

The premium price for a quicker ride has proved attractive for highway drivers at peak hours. The one to two dollars per minute that drivers are willing to spend in HOT lane tolls for saved travel time is higher than the value of their time based on wages, studies showed.[4] This makes smart roads with user fees a good alternative or supplement to gasoline taxes to fund highways. Anyway, gas taxes pay only about 10 percent of the costs of traveling on congested urban freeways, according to Information Technology and Innovation Foundation (ITIF) studies. People who dislike paying taxes are willing to pay fees for convenience.

REAL-TIME INFORMATION ON THE ROAD:
JUST WHAT YOU NEED TO KNOW

Electronic signs to alert drivers to construction, traffic delays, or emergencies have grown in use since the 1960s, more recently enhanced by wireless communications offering real-time information and remote traffic management. For example, in 2012, Inrix, a company specializing in traffic data management, equipped the New Jersey Department of Transportation (NJ-DOT) with data collection and analysis technology to evaluate traffic patterns on the state's 2,600 miles of road. A NJ-DOT official says this makes possible faster decisions than in the past, such as how best to reroute traffic after an accident, and enables posting of immediate messages on signs.[5]

As roads become smart communicators, adjusting messages to conditions, they can appear more humanly interesting. Massachusetts highways feature dynamic LED signs on the roadside that provide real-time traffic information about the miles and current estimated time to a major destination. But that's not all. State DOT officials have added catchy, imaginative messages. A safety reminder for motorists who notoriously fail to signal when changing lanes says "Use Yah Blinkah" in a mock Boston accent. The "Blinkah" idea was suggested by a staff member at a brainstorming session trying to determine how to change behavior, Secretary of Transportation Richard Davey told me. It was so well received that the agency decided to include the public in the next round of brainstorming about messages the roads could send about driving conditions, safety measures, and good driving behavior.

Digital billboards with constantly changing messages are increasingly replacing traditional billboards. In 2013, Mayor Rahm Emanuel of Chicago announced that the city would exchange land along the highway for millions of dollars in revenue from digital billboard advertising to help overcome a city budget deficit. Highway safety analysts immediately criticized the decision, citing data showing that digital signage distracts drivers longer than traditional billboards, and that nearly 80 percent of all car crashes are due to driver inattention (distraction, fatigue, or looking away).[6] Ending distracted driving is a new public health focus, led by my Harvard colleague Jay Winsten, who spearheaded the designated driver campaign to stop drunk driving.

Digital advertising is happening anyway. Digital billboard displays can rotate ads or show breaking news, one-day sales, and other short-term special events. Eventually, through the proliferation of sensors in vehicles, digital displays will be able to communicate with a vehicle to display a message custom designed for the driver, increasing the ad's value. Today, customers of large billboard owners, such as Lamar Advertising, Clear Channel Outdoors, and Daktronics, can use wireless networks to update their ads remotely and in real time from their office or home computers, as frequently as they like.

Lamar Advertising controls its digital signage for thousands of billboards across the United States at a sleek high-tech headquarters hiding in an old building in Baton Rouge, Louisiana. I toured it with CEO Sean Reilly, who pointed out the art from early billboards, such as a gigantic Barnum & Bailey's circus display, coexisting with open stairs and work spaces similar to those at high-tech companies anywhere. Surrounded by nostalgia, rows of programmers at rows of computers watch screens and shifting messages at a mini-version of an airline operations center like American Airlines' facility in Texas. Reilly had already converted 15 percent of Lamar's billboards from the old inert wooden versions to digital form.

It seems that the only limit is imagination. Perhaps someday the road can order your family's dinner automatically when you get close enough to the digital billboard for your favorite take-out place, and deduct the cost from your transponder. Less fanciful, but more socially desirable, opportunities come from a combining of sensor technologies that already exist. Econolite executive Persephone Oliver says she is particularly excited about the potential for mobile phone apps connected to sensors at intersections. This could provide alerts to the blind if a vehicle is coming toward them or they stray from crosswalks. A kind of seeing-eye phone.

Actually, imagination isn't the only limit. Deployment of an innovation like support for the blind can be slow in America, because it requires collaboration across industries and sectors. Mobile app developers, infrastructure sensor manufacturers, and champions for the visually impaired aren't generally in contact, and, to top off the complexity, roads are largely public spaces, and so that means government involvement.

Regional road innovations are not yet connected to national strategies or systems. Only recently did New England toll systems connect with New York's EZ Pass to integrate electronic tolling from the Mid-Atlantic through New England. Similarly, states are just starting to cooperate to catch interstate toll violators by using interconnected video cameras and databases. (This might not be good news for violators but is good for the rest of the public, which would otherwise subsidize them.) That these steps are so small shows how difficult it is to connect the roads that supposedly connect the nation. There is no national standard or single transponder system that is deployable across the nation. Unlike many countries, the United States does not yet have available a national smart card that can operate across all modes of transportation and all jurisdictions. The recent hacking disasters indicate that even credit and debit cards have yet to be reinvented so that they are safely usable in the digital age. The United States is just beginning to determine how network-based transportation, such as smart roads, could become a nationwide system. So far the seeing-eye phone is merely a vision.

VEHICLES: BIG DATA IN THE DRIVER'S SEAT

An estimated 900,000 front-to-rear vehicle crashes per year in the United States cause human tragedies and economic damage of about $15 billion in losses due to death, injury, and property damage, the National Highway Traffic Safety Administration tells us.[7] Technology-infused cars, trucks, and buses could potentially reduce this human and economic harm. Developers think they could also manage fuel, reduce emissions, keep you in your lane, avoid weather delays, find parking, and update your calendar.

As roads are getting smarter, so are cars. Vehicles are changing, to interact dynamically with roads and one another. Sensors, software, and networks are now allowing vehicles to communicate without human intervention in ways that reduce congestion, improve efficiency, and increase safety. Equipping cars, trucks, and buses with network connections can help vehicles avoid danger and collisions, thus saving billions of dollars and enhancing human lives. Why depend on a driver when you can depend on Big Data?

This shift is bigger than the one from manual to automatic. It makes software more important than the hardware. Cars are turning into smart robots that are big enough to move people. The essential skill for auto manufacturers is moving from metal bending to data analytics. Vehicles used to offer additional benefits by functioning as dining rooms, with cup holders a major improvement for those who use their cars to eat and drink. Today, they are turning into Wi-Fi hotspots and computer network hubs.

The fortunes of U.S. and international auto manufacturers have already changed dramatically. An industry once leading the world, and overflowing with brands, consolidated after World War II to the Big Three plus American Motors. In 1970, General Motors, Ford, and Chrysler together held an 82.1 percent share of the domestic market, and Volkswagen a barely discernible 5.6 percent, more than twice the sales of American Motors. By 1980, that combined strength had slipped to 73.8 percent, with Toyota now on the screen at 6.2 percent of the U.S. market. By 2000, the Big Three were becoming the Smaller Three, with a combined share of 64.8 percent, and Toyota was getting close to them as number 4. In November 2008, two of the three, General Motors and Chrysler (which had briefly been Daimler Chrysler) went hats in hand to Congress (but some said not humbly enough) to request assistance. Ford, under the leadership of widely admired CEO Alan Mulally, decided not to seek government money, choosing instead to overhaul its product lineup radically, simplify operations, and transform its culture, although it did benefit from federal assistance to suppliers.

In January 2009, stimulus funds were directed to GM and Chrysler. In June 2009, GM entered bankruptcy. The U.S. Treasury provided another $30 billion in return for 60 percent equity in GM, eventually selling shares to the public in IPOs in November 2010 and November 2013. At the end of 2010, Toyota was close to edging out GM and Ford at the top, as they became the Even Smaller Two; GM's market share was 18.8 percent, Ford's 16.4 percent, and Toyota's 15 percent, with Honda at 10 percent, and Nissan, Hyundai, Volkswagen, Kia, and Daimler joining Chrysler in the single digits.[8]

The industry is always adjusting, but the slow pace of major innovation is illustrated by one area users care about: fuel efficiency and carbon emissions. Hybrid-electric vehicles have been around since 1898, when Ferdinand Porsche built the Lohner Electric Chaise, but the last hybrid sold in that era was the Woods Hybrid of 1917, a commercial failure. Electric cars were also introduced in the same era, but none were commercial successes. In 1993, the Clinton administration sponsored the Partnership for a New Generation of Vehicles, to develop a clean, efficient car. Toyota was excluded from the partnership because it is a Japanese company, and Buy American sentiment was strong in that era. So Toyota launched its own secret project, G21—the global car for the twenty-first century—and beat everyone. In 1997, Toyota introduced the Prius, the world's first commercially available modern hybrid, in Japan. The Honda Insight, also Japanese, entered the U.S. market in 1999, but was quickly outsold by the Prius when it reached North America in 2000.[9] Technologies developed by the Clinton-era partnership helped U.S. automakers respond to the Prius with their own hybrid-electric cars, as well as providing engineers with deeper experience in electric propulsion. "GM met the challenge and raised the stakes with a 108 miles per gallon fuel cell vehicle," recalls Rodney Slater, former U.S. DOT secretary. Today, almost every car company with sales in the United States offers hybrids.

In 2008, Tesla Motors, a start-up company, introduced the twenty-first century's first commercially available all-electric car, the Tesla Roadster, with new kinds of batteries enabling it to run up to 200 miles between charges. Though sales were in the low thousands, Tesla turned a profit by the first quarter of 2013. Mitsubishi introduced its own electric vehicle to the Japanese market shortly afterward, in 2009, followed by the 2010 U.S. releases of GM's Chevrolet Volt and the Nissan Leaf. In 2011, Ford introduced the electric Focus, and Volvo and Smart started selling models as well. Since then, Tesla has introduced its lauded Model S, a luxury sedan, and other manufacturers have followed suit, though none to date have achieved the recognition and sales success of the Tesla. CEO Elon Musk, a South Africa–born California entrepreneur who grew rich on PayPal and other investments, feels the established auto industry is moving too slowly; in June 2014, he announced that Tesla's patents would be open for use by anyone serious about using them to reduce carbon emissions. (He's also the hyperloop visionary.)

One limiting factor is battery life and the availability of places to recharge batteries. Tesla has led the way by adding a network of charging stations for battery-powered cars and encouraging others to develop stations as well. Pressure to do this is coming from outside the auto industry. Electric Generation, a nonprofit campaign representing major utility businesses, with a nice play on words in its name, advocates for widespread adoption of electricity as a transportation fuel. Duke Energy has partnerships in North and South Carolina, Kentucky, Ohio, and Florida to conduct research on electric vehicle technology and deployment of vehicle charging stations. In October 2013, governors of eight states (California, Connecticut, Maryland, Massachusetts, New York, Oregon, Rhode Island, and Vermont) pledged to cooperate to dramatically multiply the number of zero-emission cars on the nation's roads by speeding the construction of charging stations and related infrastructure.

For all the promise of electric cars for one of the quintuple wins, environmental sustainability, this innovation by itself doesn't add-

ress other pain points caused by cars on the roads: traffic congestion, wasted time, or safety. For that we must turn to the same force
creating smarter roads: wireless networks, sensors, smartphones,
and computers that think. Software is already promising to be more
powerful than engines for the future of mobility.

THE DASH TO OWN THE DASHBOARD

What powers the car could turn out to be less important than
what controls it. The dashboard is becoming the conduit for entertainment, communications, navigation, fuel management, on-road
adjustments, emergency services, in-motion repairs, and more. Ani-
mated dinner table conversations that were once about cool new car
models are now about cool new smartphones and the apps they contain. Wireless technology networks are reinventing the car—mobility for mobility.

Mary Barra thinks the auto industry missed a bet involving technology, and as CEO of General Motors, she seems determined not to
let that happen again. She came from a GM family, went to General
Motors Institute, and rose through the ranks to be head of global
product development, the role she had when I first met her. Immediately after her promotion to CEO in late 2013, she faced escalat-
ing revelations about past safety defects, defects that caused deaths
and the recall of millions of vehicles, giving her the task of changing
the culture of the company and convincing Congress and the public
that she has. But in the brief window before her first congressional
testimony, she talked about a different topic: how the early years of
mobile connectivity were a lost opportunity for the auto industry.
"Remember that the original mobile phones were called car phones,"
she says. "That was an opportunity to seize, and automobile manufacturers didn't take it."

Barra notes that more than two-thirds of U.S. car buyers own
a smartphone and that over 80 percent of them say that connectivity influences their purchase decision. "Connectivity is key,"
she explains. But customers also want simplicity and integration,

including incorporating connectivity in the vehicle in a way that does not distract the driver from the road but rather makes the driving experience safer. This is why vehicle connectivity and sensors, processors, and memory are two major innovation priorities for GM's development efforts, alongside clean energy, advanced materials, and manufacturing technology. Ford and GM have a presence in Silicon Valley; in 2013, GM's venture capital arm invested many hundreds of millions of dollars in several dozen tech start-ups. Meanwhile, auto companies in Germany, South Korea, and Japan have been racing ahead with smart vehicles.

Is this a turning point for the industry? Connected cars are smart, coming alive the way roads are. Demonstrations show that they can sense, communicate, and operate by themselves.

SENSING: CARS THAT MAKE DECISIONS

We'll soon be relying on sensors instead of our senses. Sensors seem to beat humans in measuring the location, distance, and speed of potential obstacles on the road more reliably than a driver in most circumstances.[10] After all, the driver is only estimating; the system is calculating.

Radar was once considered the most promising sensor technology because of its reliability, low cost, and ability to detect the distance and speed of vehicles. However, while radar is useful in brake control for reducing the risk of a crash, it cannot always detect the full range of possible road obstructions. Flashing lights on cars that provide information to a driver are nothing new—consider the symbols that light up when the oil needs changing or when gas is low or when something is malfunctioning. A GPS navigation system senses a car's position and tracks it, giving directions to drivers, and speaks in a neutral voice, with no hint of blame for missing turns. Video cameras for rear views give information to drivers, bleeping ever more rapidly if objects are too close. These are all useful features, but the car doesn't do anything about a potential hazard; it waits for the driver to act.

With new digital sensor technologies, cars could instantaneously decide for themselves, processing and acting while drivers might still be reacting to a shock. The next generation of decision-making software will be able to understand the images that it sees and make sense of them to take an action, such as redirecting a vehicle or applying the brakes.

IBM's Watson—the supercomputer that made a splash on national television by beating human champions on *Jeopardy*, and was personified by a cute robotlike machine with flashing lights—is programmed to understand images and video based directly on the content of objects, not through metadata analysis of a system. In mid-2014, IBM CEO Ginni Rometty and Apple CEO Tim Cook stood together in Silicon Valley to announce a partnership to combine data analysis, cloud, and mobile technology with smartphones and tablets. Rometty told me at a quiet lunch later that transportation is an important target area and that Watson plays the leading role. Recognition of objects by computers is directly related to vehicle communications and other kinds of machine-to-machine direct interactions, making cognitive computing, or machines that think like humans, a "tremendous leap forward for transportation, especially autonomous driving situations," says Guruduth Banavar, IBM vice president for industry solutions. With Henry Ford Watson joining the team, cars can get smarter a lot faster.

The question here, as with other aspects of the smart ecosystem, is who will pay for the development, and who will align industry partners with public-sector officials behind it.

COMMUNICATING: CARS THAT INTERACT

When many people first hear the term vehicle-to-vehicle communications, they think of drivers calling other drivers on their smartphones (hands-free, of course) to tell them to get out of the way. Sorry, but that's So Yesterday.

In true V2V, as vehicle-to-vehicle communications is called, your smart car doesn't need you except to turn it on, and maybe not even then. The car can send and receive messages itself in nanoseconds.

A car is not just sensing; it is interacting, in essence talking to other vehicles and the roads and making decisions on the basis of this exchange of information. That's how smart cars can "refuse" to crash. This is helped along by vehicle-to-infrastructure communications, known as V2I, for smart roads, tunnels, bridges, or anything else that's wirelessly connected in the immediate environment. Verizon CEO Lowell McAdam says that the auto industry will drive a great deal of innovation and growth in mobile machine-to-machine communications traffic, which he sees growing at a compounded annual rate of nearly 90 percent.

Almost six million Americans own vehicles connected to emergency services through OnStar and its blue button in the car. OnStar enables sensor-based automatic crash response, remote services such as unlocking the car door, vehicle location alerts, turn-by-turn navigation, and hands-free calling, all in one system. Along with its recent affiliate in China, OnStar is a wholly owned subsidiary of GM serving more than seven million customers worldwide, 90 percent of them in North America; in 2013, it was available on thirty-nine GM-branded models, backed up by emergency response teams.

In 1993, engineers from Electronic Data Systems (EDS), then owned by GM, and telecom Pacific Bell, now part of AT&T, approached GM with an idea for a microwave-based mobile two-way voice and data communications system for a vehicle. Although then financially distressed GM declined, the concept evolved through informal collaboration, and emerging cellular phone technology replaced microwave. In 1994, as GM's finances improved, a joint venture between GM North American Vehicle Operations, Hughes Network Systems, and EDS was formed to develop the product eventually named OnStar. In 1996, its first year as a subscription-based service, OnStar had 1,100 subscribers, growing to 100,000 by the new millennium. In 2001, OnStar reached one million subscribers. Ford's competing system had started just prior to OnStar but dissolved by 2002. By the end of the decade, GPS navigation was available for nearly every manufacturer, but there were only a few full-service OnStar competitors, such as BMW Assist and Mercedes-Benz Tele Aid. In 2008, GM's Chevrolet Malibu was marketed with OnStar as a big piece of the offer, contributing to a 50 percent vehicle sales rate increase compared with the prior three-year average. Yet in 2009, when over 5 million car owners subscribed, nearly half of GM customers did not activate or renew their factory-installed OnStar systems, a sign that something was still missing.

Connection can't be fostered in isolation. Seems obvious, but this was a dramatic change in mindset for large manufacturers. A GM-proprietary system, which doesn't articulate with the architecture of other auto systems, remains limited without an entire ecosystem around it. It's like an Internet without email or websites, or a smartphone without apps. It's possible to make one unique car at a time to be driven independently by one driver at a time, but for vehicles to be connected communication centers, companies and development efforts must be connected too.

That was Mary Chan's job. From May 2012 through late 2014, she was the face behind blue OnStar buttons in GM cars, as president of GM's Global Connected Consumer, which gave her the responsibility for expanding and enhancing OnStar and other connected vehicle services. Named a Top Woman in Wireless and on a list of Ten Mobile Game-Changers, Chan spent twenty-five years working in the wireless infrastructure group for Bell Labs and Lucent. Her last job with Alcatel Lucent was to develop 4G LTE, the latest, fastest wireless services, for Verizon and AT&T. She left in 2009 for Dell, where she worked on enterprise mobility products and services. GM recruited her in May 2012 to take OnStar to the next level, tapping the full potential of wireless networks. When we spoke close to her two-year anniversary, she explained that she took the job because she felt that GM might be making the biggest global commitment of any auto manufacturer to wireless technology.

Chan says her mission is to bring vehicles into owners' digital lives and their digital lives into their vehicles. With most 2015 models in the United States and Canada equipped with a built-in 4G LTE Wi-Fi hot spot, there are benefits for multiple users within a car. The system allows up to seven devices, such as smartphones and tablets, to connect to high-speed wireless Internet in and around the vehicle. The built-in hotspot provides a better in-vehicle experience, Chan says, than a smartphone's hotspot, with a stronger signal and a fast, reliable connection. Simultaneous voice and data let a driver talk to an OnStar adviser while the car talks data for a diagnostics report.

It can be educational for everyone. The driver can understand the performance of the vehicle and be guided away from harm. School kids can access special content on family-friendly learning apps Famigo and TumbleBooks, eBooks for kids. This is made possible by the development of 4G wireless network systems, which are significantly faster and more capable than 3G networks. Such systems, with data in the cloud, become platforms that are less dependent on what's built into the vehicle. A connected car permits updating over time, as the customer uses the vehicle. Communications capabilities can be serviced or augmented remotely without bringing the car to a dealer. To do all this, "wireless, consumer electronics, and the auto industry must come together," Chan declares.

A billion cars are on the road worldwide, and a billion mobile devices use Google's Android software, but those large groups are barely connected. The Open Automotive Alliance aims to change that. Launched in January 2014 with GM and Google as founding members along with Audi, Honda, Hyundai, and graphics interface giant NVIDIA, it is a global alliance of technology and auto industry leaders committed to rapid development of Google's Android Auto as a common platform. In June 2014, another two dozen companies joined, including competitors Chrysler, Ford, Mazda, Nissan, and Volkswagen, and technology companies Freescale Semiconductor, Fujitsu, Mitsubishi, and Panasonic. This is "an open ecosystem for the open road," the alliance proclaims.

Public-sector transportation officials also see benefits in V2V communications for safety, fuel efficiency, and reduced congestion (for example, potential spacing of cars). They're putting feet on the accelerator to gather information on results, hoping to speed creation of a legal and policy framework for promising technologies.

The Volpe National Transportation Systems Center, near MIT in Cambridge, Massachusetts, part of the U.S. Department of Transportation, began by enlisting eight major auto companies in a multifaceted research collaboration with Volpe, Michigan DOT, and the University of Michigan's Transportation Research Institute that would support nationwide expansion of V2V communication technologies. Volpe director Robert Johns and key staff Anne Aylward and Ellen Bell described this to us during a September 2013 visit. Each manufacturer built 8 pilot vehicles, for a total of 64 vehicles equipped with a full suite of smart sensor technology. The cars were tested for three months with sensor technology and three months without. Later, an additional 150 vehicles with a full range of after-market network safety systems, such as crash avoidance, lane departure warnings, and emergency brake light detection were to be added midway through the studies. Johns and his team concluded that early results strongly support the use of network technologies to improve road safety.

For all the promise, GM CEO Barra sounds an obvious cautionary note: "V2V works when V2V exists." That time is not quite yet. Although vehicle communication technology is well developed, integration of automobiles and infrastructure with that technology is still a work in progress. Barra urges leadership. "We just have to make the decision to do it," she says. "The coordination has to happen, and there has to be a clear vision." Cross-industry, cross-sector collaboration is essential: auto manufacturers at the table with software companies, sensor makers, wireless network operators, and public officials.

CONTROLLING: CARS THAT DRIVE THEMSELVES

Sensing and communicating combine to make possible a third, more futuristic, smart-car capability: full control. By using sensors and networks, autonomous cars don't need drivers at all. The prospects are a children's toy fantasy come to life, but the toy cars are controlled by radio devices over short distances. Drivers will control their driverless cars with smartphones over wireless networks. We could send our autonomous vehicles to find parking, or summon our cars when needed from remote locations, to arrive exactly where we are, thus saving time in traffic and reducing traffic congestion. Hands-free driving could move from occasional cruise control to long stretches of time that could be used more productively than by watching the road.

Automated cars are ready to go and soon could be big business. Some industry analysts expect driverless cars to be sold commercially within the next ten years at a price about $10,000 more than driver-controlled models. It is expected that the proliferation of driverless vehicles will cut insurance costs because of reduced risk of accidents, and will improve overall fuel efficiency. According to Morgan Stanley, autonomous vehicles have the potential to contribute $1.3 trillion in annual savings in the United States.[11] ITIF president Robert Atkinson puts the savings at $1 trillion and argues that this size of contribution to the economy makes a case for tax credits for autonomous vehicles. Of course, first the optimistic forecasts must materialize.

The centrality of software to America's, and the world's, mobility future is one reason Google is in the transportation business. Google is testing driverless cars with Google software on select roads in California. But German manufacturers are moving fast. BMW and Audi showcased their latest autonomous electric vehicle models at the January 2014 Consumer Electronics Show in Las Vegas.[12] Mercedes-Benz's self-driving S-class sedan began to be tested on German city streets in September 2013, thanks to a European regulatory environment more conducive to autonomous vehicles, already allowing automated steering up to 6.2 miles per hour, and corrective steering above that, as long as drivers can override it. The United States lags in updating regulations to allow use of autonomous vehicles. Only California, Nevada, and Florida have permitted autonomous vehicles to operate on public roads, and always under strict guidelines.[13] Some attribute federal foot-dragging to security concerns, that autonomous vehicles could be strapped with explosives and sent into a public area; proponents see that as an excuse, since conventional vehicles could do similar damage. But there is also a reservoir of public skepticism. Many people don't trust other drivers, and it is not clear whether they would have more or less trust for vehicles without them. They might not want an autonomous car pulling up next to them, let alone malfunctioning and getting into an accident.

Whereas U.S. federal regulators say that fully autonomous vehicles aren't ready to operate on public highways except for testing purposes, experts argue that regulators could allow incremental innovation—for instance, use of driverless cars in low-risk, controlled environments, such as exclusive lanes for autonomous buses or trucks. But then local and state officials would have to smarten those lanes. To deploy autonomous vehicles nationally, federal and local regulations will have to accommodate both V2V and V2I communication.[14] Although industry and government are cooperating on the joint research I've just described aimed at deploying V2V, that is only in a scenario where cars are still controlled by drivers. To maximize the potential of driverless cars could require more smart roads. Building a regulatory framework for an autonomous driving environment will also be a challenge since roads and highways are regulated at the local, state, and federal levels, and all must agree.

Sensors and communications seem simpler by comparison. They empower people without fully replacing them. Clearly, these are debates that the public should get involved in, and not simply let their cars do the talking. Public attitudes about mobility are already changing, especially as new smartphone-based services are making us mobile without owning our own cars.

USERS: APPS AND SHARED SERVICES

Driverless cars, meet carless drivers. People who want personalized transportation have an increasing array of services and business models to choose from, thanks to smartphones and mobile networks. They don't even have to own a car. In an annual survey conducted by Zipcar, millennials consistently and overwhelmingly say they would give up a car before they would give up their smartphones. Why own, when there are other ways to get there? Now they can share cars or share rides, without parking hassles, especially if they use apps to find parking.

Vehicle- and ride-sharing programs are examples of new business models that deploy information and communication technologies to optimize vehicle usage, which means things like fewer cars sitting in garages or on the street, unused most of the time. A rising social movement promotes less consumption, pushing a "sharing economy," as it's become known, with "collaborative consumption," another movement name, especially appealing to those concerned about not only the environment but also saving money. At the same time, new technologies open new possibilities. Users see a convenience for themselves, sometimes at lower cost and with some expression of values thrown in. Entrepreneurs see a hot new business opportunity. Public officials in cities see a chance to ease congestion problems. All brought to you by sensors, smartphones, and Big Data.

Mobile app-based vehicle- and ride-sharing programs are helping to reduce the number of vehicles on the road. Surveys of commuters in some major urban areas show that up to a quarter of car owners sold their vehicles because of the convenience of vehicle- and ride-sharing services. Studies find that for every vehicle added to vehicle- and ride-sharing fleets, an average of five personally owned vehicles come off the road, with some variation depending on the city surveyed.[15] Fewer cars, along with rides on demand, bring us closer to the quintuple wins: less congestion, fewer safety hazards, convenience and time-saving, and cleaner air with lower carbon emissions. That's why new companies such as Uber have taken on so much value—but not without controversy.

RIDE SHARING: UBER AND FRIENDS

"I think I've got 20,000 years of jail time in front of me," Uber founder and CEO Travis Kalanick joked when asked at a Tech-

Crunch conference in 2011 about Uber's highly publicized legal challenges (one reason for changing the company name from UberCab). His calculation came from a threatened ninety days of jail time by the City of San Francisco for every ride they offered, plus a fine of several thousand dollars.[16]

Cease-and-desist orders against Uber, Lyft, and SideCar were removed by the California Public Utilities Commission in 2013. But consider the irony of this situation: progressive, tech-savvy, environmentally conscious, entrepreneurial San Francisco was in the forefront of blocking a progressive, tech-savvy, environmentally conscious, entrepreneurial company. And equally innovation-conscious New York City and Cambridge, Massachusetts, are in the same battles. This says a great deal about the challenges of change, when upstarts threaten establishments and upset the previously negotiated set of rules.

Uber, founded in 2009 and first launched in San Francisco in summer 2010, concluded a funding round in late 2014 that valued the company at a whopping $41 billion. Kalanick claims that the company (still private) has doubled revenues every six months since its founding.

Uber provides on-demand car services to consumers via apps for iPhones and Android smartphones. Using proprietary algorithms, the app connects customers to drivers in its network, who also operate the app, and provides cost estimates, estimated time of arrival, and driver reviews. Once a trip is complete, customers pay through the app, which connects to their credit cards—tipping drivers with cash is prohibited, and tips are included in the electronically charged fare. Uber has charged drivers 20 percent of the trip costs, with commissions falling as competition has increased. Drivers and riders rate each other.

For its first two years, Uber focused on black-car (livery or limousine) service, an industry in which as much as 50 percent of a driver's workday is downtime. Uber promised to optimize use of under-utilized black cars by allowing off-duty drivers to find additional customers at a slight discount to normal black-car service rates; customers could select regular town cars or SUVs. In the years since, however, the range of available services has grown exponentially. In spring 2012, the company rolled out Uber Taxi, bringing cabdrivers into its network, and that summer introduced UberX—a low-cost ride-sharing service, priced at a slight premium to cabs, that allows private vehicle owners to offer rides. In January, the company slashed prices for UberX to undercut cabs and encourage wider adoption of the service. More recently, the company has also offered vehicles with car seats for children or large cars for big groups. In the works are delivery services such as Uber Rush, a service piloted in Manhattan. Industry spillovers are all over the place.

Uber owns technology, not cars, and uses it to full advantage. Using predictive analytics and a number of data points, the company also practices dynamic pricing in its services. During periods of high demand—a Friday evening or the end of a big sports event—Uber's "surge pricing" charges a demand-based multiple of its base fare. Uber has been criticized for implementing surge pricing during actual storm surges; for example, in New York during Hurricane Sandy in 2012. A bad idea.

We're still sorting out exactly how to categorize these new business models. Claims of unfair competition (driving out taxis that paid for medallions) or operating an unregulated service (how safe are the drivers?) have echoed throughout reactions to such services. For example, is Lyft an example of peer-to-peer ride sharing, or is it a for-hire livery service using a mobile app? That is the question in the lawsuit filed against Lyft by New York State just two days before its launch in New York City. Lyft, founded in San Francisco a few blocks away from Uber and a few years later in 2013, has thirty cities to Uber's forty-five as of mid-2014, charges drivers less, and carries the tagline "your friend with a car." It originated as a service from Zimride, a nationwide ride-sharing service largely for intercity trips started in 2007 and sold to Enterprise in 2013 so Lyft could focus on the urban market. Its surge pricing is called Prime Time and its bargain times called Happy Hour. But does all that make it a peer-to-peer community? Or does the other California competitor Side-Car's decision to let drivers set their own prices bring it back from the brink of the purely commercial to the ride-sharing realm?

Insurance companies also want to know. And they want to know the outcome of local regulatory negotiations. A bill passed by the California Senate in September 2014 requires ride-sharing companies to insure drivers from the minute they sign on, like taxi drivers, instead of the minute they begin a ride. But we're back to definitional confusion. These are part-time independent contractors using a little extra time in their own cars. Will the California bill entail insuring whole swaths of San Francisco?

These services are growing quickly worldwide as well as in the United States because of rising consumer demand and satisfaction. There is too much momentum to stop now. This is one field that could be left to private markets to shape—with the exception of safety, which is clearly a public-sector consideration.

CAR SHARING: ZIPCAR AND HOW TO ZIP AROUND TOWN

If you want to drive yourself, you can still participate in the sharing economy. That's the philosophy that led to Zipcar. It's another example of how big commercial propositions often emerge from idealistic, community-oriented founders, like Pierre Omidyar of eBay, who started the online auction giant to help his girlfriend trade Pez dispensers and help others become a community to exchange goods, and later became a billionaire as eBay grew its business.

Zipcar's founding story is appealing. Before the dot-com crash, well before the smartphone, Zipcar was launched by two educated stay-at-home moms spending sixty-seven dollars. Robin Chase and Antje Danielson were environmentalists who wanted to reduce the number of cars on the roads. In 1999, there were few models for how to do this, a Mobility Cooperative in Switzerland and Flexcar, a car-sharing service in Portland, Oregon (which merged with Zipcar in 2007)—and that was about it. But the founders lived in innovation-conscious Cambridge, Massachusetts, and were connected with MIT and Harvard. Zipcar was a technology pioneer, using remote wireless networks to permit access to cars unattended, with no key-swapping required. It grew in a dozen years to a business valued at nearly half a billion dollars.

Danielson was gone in 2001 and Chase in 2003 when Scott Griffith replaced her as CEO; Chase was named to *Time*'s list of 100 most influential people in 2009 and remains active in the car-sharing world. Zipcar wanted to accelerate growth with next-stage leadership; the board considered the ability to raise capital as part of its new CEO pick. Griffith, who had worked at Boeing and a large consulting firm, saw a giant opportunity in the little start-up, even though it was bleeding cash. It hit the sweet spot of his interests in technology, transportation, and cities. He saw that this political movement, the sharing economy or collaborative consumption, could produce a big company, and it did. Zipcar operates in 196 U.S. cities and 24 international markets. Griffith stepped aside as CEO in 2013, several months after the car rental company Avis Budget Group paid $491 million to acquire Zipcar.

In April 2014, shortly after he spoke at the Harvard summit, I caught up with Griffith in his new offices at General Catalyst, a venture capital firm in Harvard Square in Cambridge, the day after his affiliation was announced publicly. I knew him throughout the Zipcar years from serving on a board together. Now he's looking for growth-stage investments, cofounding a company that's developing a smartphone app for auto insurance; drivers could connect to have their driving monitored and potentially get cheaper insurance.

Zipcar was not immune to the political battles with local officials that ride-sharing companies faced, but the focus was a little different. Whereas ride-sharing involves no physical inventory, car-sharing requires cars and, thus, places to put them. For Zipcar to succeed, cars must be near users, so people can just hop in and go. Around 2004–05, Griffith wanted Zipcars curbside, not only in selected parking lots and garages. In Washington, D.C., he went around and around on that for

years with the city council, he recalled, with small shopkeepers object-
ing to the prospect of losing spaces in front of their stores. Finally
there was a hearing in which he said that half the city's population
owned a car, and half didn't. "'You are allocating 100 percent of the
spaces to half the people. Just allocate 1 percent to car sharing. We'll
pay market rates. If it costs $200 per month in a garage, we'll pay that.'
The meter brings in only $75 a month, so they'd make money on it," he
told me. About two years later, the city council authorized a small pilot
project. As Zipcar's partnership with the city grew, so did significant
incremental parking revenue.

Griffith feels that Zipcar makes life easier and cheaper in cities.
He cites figures showing that use of Zipcar could reduce the 19 percent of the average family of four's budget spent on transportation to
a mere 5 percent. The 19 percent figure is from the Federal Highway
Administration in 2011; a subsequent analysis said the share of those
families' budgets had dropped to 11 percent by 2014.[17] One calculation notes that the $8,876 annual cost of owning a car in Chicago (an
American Automobile Association estimate) would cover 109 full
days of driving at Zipcar's highest weekend rental rates.[18] Drive less
than every weekend, and save considerably.

Griffith still loves Zipcar, but he worries about the culture and
brand, a common concern following acquisitions. I am very familiar
with the issues: established companies buy upstarts but don't always
know what to do with them. They often veer to extremes, either swallowing them whole with no traces left, thereby destroying value; or
keeping them off to the side unintegrated, thereby failing to get cost
savings, brand synergies, or learning benefits.

Some established companies are doing a better job of adding car
sharing to their portfolios, as well as technology to their operations.
Two in particular: Enterprise, in car rental services, and Car2Go,
the offshoot of a German auto-manufacturing company. Each comes
from a different place and a different industry, yet intersects with a
similar service.

Enterprise Holdings in St. Louis, the car rental industry's largest player, was a bidder for Zipcar. Chairman Andy Taylor's father founded the privately held company; Pamela Nicholson, a thirty-two-year company veteran and long-tenured COO, was named CEO in mid-2013 as the first outsider in that post. Enterprise's 2013 revenue totaled $16.4 billion, and it operates approximately 1.4 million rental vehicles at 8,100 locations and employs 78,000 people. In addition to the flagship Enterprise Rent-A-Car brand, Enterprise Holdings owns Alamo and National car rental services, a business fleet management unit (Enterprise Fleet Management), and a truck rental service, Commercial Trucks, which competes with U-Haul. It expanded into ride sharing by buying Zimride from Lyft, and starting Enterprise Rideshare, a vanpooling and carpooling service that partners with companies, governments, and individuals to provide pooled rides. Enterprise has added self-serve electronic kiosks and systems permitting unattended vehicle access. Through Enterprise, customers can rent, buy, or share cars.

Enterprise formally launched Enterprise CarShare in spring 2013, but experiments with car sharing within the company date back to 2005. Prior to the 2013 launch, Enterprise partnered with universities, government, and business to provide local car sharing, and made a series of small acquisitions: WeCar in St. Louis, Philadelphia's PhillyCarShare, IGO CarSharing in Chicago, Mint Cars-On-Demand in New York, and most recently AutoShare CarSharing in Toronto, in March 2014. Enterprise CarShare operates in the United States, the UK, and Canada in partnership with eighty universities and forty government and business campuses. The service heralds a focus on sustainability, sponsoring Earth Day events showing carbon footprint benefits of car sharing. A partnership with the Salt Lake City municipal government provides free metered street parking to members using hybrid cars within the city.

Enterprise CarShare has grown to become, along with Hertz 24/7, Zipcar's primary direct competitor, though with less scale and name recognition. But Scott Griffith expresses considerable admiration for Andy Taylor's and Enterprise's future.

Another competitor drives in from a different direction—German giant Daimler AG, maker of Mercedes. Some say that Daimler turned a problem—a line of small Smart cars that didn't sell—into an opportunity, using those cars as the basis for a car-sharing service, Car2Go.

Car2Go is a "free-floating" car-sharing provider, a little different from Zipcar, offering one-way, point-to-point rentals accessed on demand via smartphones. The service is priced to compete with both hired rides, like taxis or Uber, and other car-sharing services like Zipcar, which announced a competing point-to-point service in 2014. Car2Go utilizes exclusively Mercedes-Benz Smart Fortwo vehicles, produced by Daimler, and charges by the minute, hourly, or daily, with an emphasis on urban commuting. In most of its markets, the service partners with municipalities to offer free on-street parking. In Washington, D.C., for example, one of its major U.S. markets, Car2Go vehicles may be parked in any city metered parking spot for free, and users may end their reservation anywhere within the metro area. Since the cars all have Internet-connected GPS sensors, available locations are displayed live to users, and typically cars taken in an area tend to be replaced by other trips ending in the same location. The onboard computer in the car provides free navigation and, emphasizing the environmental impetus for car sharing, analyzes each trip to provide drivers with an "eco-score."

The idea sprang from an internal innovation unit and began in Ulm, Germany, for Daimler employees. Car2Go entered the United States with a rollout in Austin, Texas, in May 2010, followed by an expansion in 2011 to other European markets, Vancouver, and San Diego, and then by a 2012 U.S. expansion to Washington, D.C., Portland, Oregon, Miami, Seattle, Denver, Minneapolis–St. Paul, and Columbus, Ohio. By 2014, Car2Go operated in ten U.S. and nineteen international markets with about 10,000 vehicles and 700,000 subscribers, hoping to reach a million subscribers by the end of fiscal 2014. Seattle and Washington, D.C., are the service's largest U.S. markets by membership and fleet size, with 35,000 members/500 cars and 30,000 members/400 cars, respectively, with Austin a close third, though dwarfed by the 1,200 cars in Berlin. In a sign of what could be coming to your neighborhood, the service is adding electric cars, hybrids, and Car2Go Black featuring Mercedes B-class cars, which might show off the Mercedes, and perhaps eventually help sell a few cars too.

Car2Go is housed in a new business unit, Daimler Mobility Services, established in January 2013. The unit has grown to include other mobility concepts like bus rapid transit, which I'll describe in the next chapter, and it's seeking start-ups to acquire, as we'll soon see.

As Zipcar's studies show, car sharing not only takes cars off city streets; it also saves time and money. If an average family's household budget used for transportation, which is largely automobile-owning expenses, could drop by half or more by the use of shared cars, that's meaningful for most Americans. After having been told for decades that owning a car is a sign of success, Americans could also be encouraged to see that there is cultural and well as economic value in shared use, just as auto industry giants like Daimler are beginning to shift to see themselves as providing mobility services rather than merely making products.

SMART PARKING: IT'S ABOUT TIME

Parking apps bear superficial similarities to the collaborative consumption/sharing economy claims made by car sharing and ride sharing. Parking apps use wireless networks and smartphones, build membership "communities," and have an underlying improve-the-environment rationale. But there's one fundamental difference: parking is less about sharing than about grabbing.

Bostonians know the grabbing scene well. It's a long tradition during Arctic winters to shovel out the parking space in front of one's home or storefront and then put lawn chairs or other large objects to mark the space and prevent others from using it. The city winks at this. But it draws the line at the new app Haystack, founded by twenty-four-year-old Eric Meyer, which allows members to notify others when a street parking space is available and hold it for them. Similar services in San Francisco were greeted by cease-and-desist orders from the city. Public officials argue that these services are benefiting from a public asset and peddling services they don't own; Meyer says he is selling information, not physical property. Meanwhile, why shouldn't the city do this itself? Boston launched its Parker app in December 2013, using sensors to show available spaces in the congested downtown waterfront Innovation District.[19] City officials say their interest is a public service (the fees are nice too) that can reduce the pollution, congestion, and lost productivity as drivers circle around and around looking for scarce parking. (Haystack is now out of business.)

Welcome to the brave new world of smart parking. It's a competitive street fight, but one with big potential for the quintuple wins. Studies estimate that 30 percent of city driving is looking for parking spots, wasting time and fuel, and causing congestion and pollution from the idling. In Brooklyn, New Yorkers say, 45 percent of fuel consumption is caused by drivers looking for parking spaces. Traditional parking systems haven't been upgraded to keep up with demand. Parking is a $25 billion industry in the United States, but significantly inefficient. Nationally there are about four times as many parking spots as vehicles, but that includes suburban and rural areas. In urban areas, the ratio of parking spots to cars is much lower, and prices are high. Just ask anyone who tries to park in Manhattan, which can cost as much per day as a hotel room elsewhere.

Smart parking systems connect parking space providers and drivers seeking them over mobile networks. Drivers use mobile apps downloaded onto smartphones to conveniently search for and pay for parking spaces, while parking providers manage their parking spaces more efficiently and adjust pricing dynamically on the basis of demand. You can even rent out your own driveway at home via an app. In this ecosystem, sensors embedded in parking spaces, lots, and garages are linked over a mobile network to drivers' mobile phones, which also serve as sensors. Verizon CEO Lowell McAdam points to smart coatings that can be used to embed chips in parking lot striping to communicate with vehicles. Smart parking apps can give drivers several parking options that evaluate costs, travel time, payment methods, and other practical items. If drivers find parking faster, they're off the road sooner. Providers of smart parking services benefit from customers' willingness to pay premiums for parking options that are highly sought after.

As highway usage evolves, parking systems must be upgraded to keep up with demand. In the United States, start-up businesses such as Streetline, ParkMe, Park Assist, ParkingCarma, ParkMobile, and Parking Panda are working with providers of both municipal on-street parking and private off-street parking, nationwide and internationally, on smart parking services. For example, Streetline, which has a partnership with Verizon, collects and organizes real-time data on parking in select urban areas in California, Washington, D.C., New Jersey, New York, North Carolina, Texas, and Utah to provide integrated parking management technology, including street-level sensors, mobile apps, and data analytics software. Streetline's improvements to municipal parking have stimulated economic growth for many cities, including Los Angeles, which is using the app to improve the parking situation and increase city revenue through dynamic pricing, charging more at times of high demand.[20]

There are numerous variations and specializations. Parking Panda provides a similar service, but focuses on parking around convention centers and event spaces, such as the Jacob Javits Center in New York and the New Orleans Superdome.[21] Parking Panda users can search for a convention center by name, compare prices of available parking in the vicinity, and reserve the spot in advance. ParkMobile, which is used in over five hundred cities worldwide, has branded its mobile app as a payment service for on-street and private parking fees.[22]

Over time, these services are collecting and storing user information that will help drivers not only conveniently find and pay for their parking but also park more efficiently in the future. Analysis of historical data will benefit parking providers by indicating which dates and times have peak demand for parking and during which they can charge premium rates. SFPark in San Francisco, which was awarded one of twenty-five Innovations in American Government Awards by Harvard Kennedy School, works with the San Francisco Department of Transportation to track open parking spots and set prices dynamically according to availability and demand. A mobile phone app directs drivers to available spots efficiently and allows them to pay meter fees remotely from their smartphone. The program has been reported to reduce double parking, noise pollution, excessive carbon emissions, and traffic congestion, removing several pain points by being smart.

Just like those in the car-sharing industry, the giants of the parking industry are not standing by idly and waiting for the industry to change. "Parking as an industry is undergoing immense technological change, and the good companies are investing in technology so they can stay competitive," Alan Lazowski, CEO of LAZ Parking, told us.

LAZ Parking was founded in 1981 as the merger of two childhood friends' small businesses in Boston and Hartford, Connecticut. A third childhood friend also ran a parking company on the West Coast, which merged with LAZ in 2007. The three friends—Lazowski, Jeff Karp, and Michael Harth—are now chairman/CEO, president, and chief culture officer, respectively. The chief culture officer shows that parking lots have a soul; LAZ promotes from within and gives to the Special Olympics, St. Jude's Children's Hospital, Komen breast cancer charity, and Make-a-Wish Foundation. Following a 2006 contract with the City of Chicago's Millennium Parking Garages, LAZ expanded beyond the East Coast and in 2007 received a strategic investment for 50 percent equity by the dominant global player, VINCI Park. LAZ Parking operates in twenty-six states and 250 cities—a total of 1,850 locations and 740,000 parking spots—and employs 7,300 people, with $830 million in 2013 revenues.

LAZ Parking is working with industry partners on a system for smartphones, which will flash green or red depending on whether

spaces are open, and can be integrated into in-car navigation systems. The system will display real-time availability of space at parking lots and garages and enable drivers to reserve available spots.
LAZ is also working with 3M to develop a payment system that
scans cars' license plates—minimizing the need for attendants and
reducing congestion at entrances/exits to garages; LAZ's technology already allows bar code–based payment. The company is also
exploring real-time dynamic pricing options for its parking.

COLLISIONS AND CONSOLIDATION IN THE APP WORLD

Sharing-economy technology-based companies provide the kind
of David and Goliath story Americans love: feisty start-ups empower
users and challenge long-established fuddy-duddy giants. That crash
you hear is the sound of industries colliding. Boundaries are blurring
between rental cars, limo services, taxi cabs, college bulletin boards,
social media communities, parking lots, and parking meters, as they
get connected. Things are happening so quickly, and so recently,
that it's hard to sort out who will dominate the car- and ride-sharing
space. But it seems certain that car and ride sharing are here to stay,
as mobility, not metal, becomes the goal.

With so much fragmentation and so many niche players, it is inevitable that super-apps would spring up to aggregate and integrate all
the other apps. RideScout is a data aggregator app for smartphones
that provides individuals with real-time information about transportation availability, cost, and speed. All kinds of transportation,
all at once: traditional public transit; ride-hiring services from traditional taxis, limos and airport shuttles, to new services like Lyft
or Sidecar; car-sharing services like Zipcar, Car2Go, or more traditional car rentals; ride-sharing services such as Amigo, Casual
Commuters, or Carma Carpooling; and directions for personal cars,
parking, walking, and biking, including finding bike-sharing services. There is also an option for anyone driving his or her personal
car to offer rides to others.

RideScout's analytics quickly integrate and normalize disparate data streams from many different transportation providers. Say you want to go from point A to point B. The app will calculate travel options and display them in a unified screen, with anticipated time and cost. For walking and biking, it adds calories burned. The advantages and disadvantages of all the possibilities will be clear to users. RideScout is available in Austin, Texas (its home market), Washington, D.C., San Francisco, and Boston, as of this writing. In Austin and D.C., RideScout has almost all the major players signed up, except for Uber, which has yet to commit. It's not hard to see why. If RideScout succeeds, could it drive Uber out of business?

Founder and CEO Joseph Kopser is not a young rebel. He's in his forties, based in Austin, and recently retired as a lieutenant colonel from the U.S. Army, where he worked at the Pentagon under General Casey. He is also active in efforts like a "Patriot Boot Camp" to help veterans find jobs in the tech sector. Perhaps because of this military background, he seems eager to partner with government in delivering transportation innovations—a big shift from the institution-averse tech entrepreneurs one usually sees. We spoke with him in June 2014, the day after RideScout was recognized by U.S. Transportation Secretary Anthony Foxx as the winner of a DOT Data Innovation Challenge. Winning comes with the opportunity to brief senior DOT and White House officials. As he talked on his mobile, Kopser was walking from meetings at the Old Executive Office Building next to the White House.

For Kopser, dynamic ride sharing is a holy grail for transportation and, in his view, the future. He notes that 76 percent of cars on the road have only one driver and suggests that widespread adoption of ride sharing would lead to at least a 20 percent drop in roadway traffic. The key to success in dynamic ride sharing, he thinks, is not just cost and speed but eliminating hassle, making it easy for users to compare and choose options. Kopser aims first to create a level of comfort with ride sharing for urban millennials and then to expand across society from there. The ride-sharing market is young and fragmented. Kopser hopes to move further, by partnering with large employers trying to improve commuting for their employees, which means RideScout functions as a ride-sharing provider; it also sells data to service providers on patterns of use.

Kopser describes himself as first and foremost a patriot; he speaks passionately about the effectiveness of public-private partnerships (in the broadest sense of the partnership ideal) when they successfully align incentives. He strongly supports setting standards in transportation—for example, streamlined ticketing processes or data standardization—and then letting the free market decide how to utilize transportation resources. In our conversation, he repeatedly mentioned a desire to commoditize transportation via tools like the RideScout marketplace and hinted that his company has thought ahead to what comes next after vigorous competition shakes out the marketplace, which, he says, is closer than you think. Sure enough, RideScout was bought by Car2Go's parent, Daimler Mobility, in September 2014.

New service models compete with established companies but also clash with the public sector, as some tech entrepreneurs don't see why they can't smash any barrier or ignore rules they consider silly. But others, such as Joseph Kopser, are willing to collaborate with the government from the beginning. Transportation officials are often ready to work with them. Massachusetts Secretary of Transportation Richard Davey wants the public sector to engage techies. We were talking in the State House while waiting for a meeting of Governor Deval Patrick's Council of Economic Advisors. The governor's three unwavering priorities during eight years in office were education, jobs, and infrastructure. Technology is at the center of all three. Davey and MassDOT have encouraged new technologies that provide information, as RideScout does, spawning over fifty smartphone apps to help customers optimize their use of transit. He often speaks on programs with technology-in-transportation entrepreneurs because organizers hope there will be fireworks. Instead, the state has encouraged Uber and its counterparts by offering them data sharing and support. (But they're on their own for parking spaces.)

WIRELESS NETWORKS: THE POWER
BEHIND THE OPPORTUNITIES

Underlying the evolving, dynamic, smarter transportation system is the infrastructure of wireless communications networks. Service apps, connected vehicles, and many other innovations are made possible by communications highways that allow smartphones to find information, sensors to reveal what they see, Big Data to move, and machines to talk to one another—the so-called Internet of Things. Communication networks empower users with information and choices, and make the automatic adjustments that get people and things to the right places at the right times, safely. To consumers, smartphones seem the most empowering of them all. In 2002, approximately one billion mobile handheld devices were in use worldwide, but they weren't yet smart. By 2010, three years after the launch of the Apple iPhone, that number had risen to five billion; and in 2015, over six billion mobile devices are expected to be in use.[23]

There's more to the network than consumer apps. Sophisticated information and communication technology can guide the operation of transportation systems with multiple vehicles, and can enable transportation service providers to gather and analyze large volumes of data about passengers, cargo, and travel and shipping patterns that further empower them to make more-intelligent decisions about their services.

The networks that connect and control everything emanate from the offspring of traditional telecommunications providers. Verizon Communications is decades and generations away from Ma Bell, the product of her breakup and several waves of mergers. Use of wired landlines has decreased; the wireless side is the growth engine. Data transmission (including streaming entertainment) has long surpassed simple voice-to-voice telecom. Verizon was not the original service provider for the iPhone but came in strong when it partnered with Google and Android, surpassing iPhone sales, ushering in a period of accelerated transformation of the company, and proving my adage that if you can't do it right the first time, do it better the second time.

OPENING DOORS TO INNOVATION: VERIZON
INVITES OTHER COMPANIES TO EXPLORE

Verizon has gone from being a walled garden, its metaphor for a closed system, to having an open door to partnerships. This is the same change of mindset that General Motors and other long-established companies must make to embrace new technology quickly.

Through its Innovation Centers in a Boston suburb and San Francisco, Verizon invites numerous other companies, from start-ups to giants, to use Verizon's facilities to develop applications that can run on the latest high-speed network, called 4G LTE (4th generation, long-term evolution) at no cost and with no obligation. The 4G mobile technical standard allows for download speeds of up to 100 megabytes per second. Verizon Wireless was among the first telecommunications firms to select LTE as its technology foundation of choice for 4G mobile service, because LTE offers peak download speeds comparable to fixed-line broadband plus the ability to operate at various frequencies of spectrum, rendering it more flexible than other 4G alternatives such as WiMAX.[24] If run in the 700 MHz (megahertz) band of spectrum, signals can penetrate buildings and walls easily and cover larger geographic areas with less infrastructure, compared with frequencies in higher bands.[25] Sounds like Supermobile, leaping tall buildings with a powerful signal.

The benefits of this technical capacity come to life in the showroom at the Verizon Innovation Center in Waltham, Massachusetts, outside of Boston, which is big sister to the San Francisco facility. In creating a new way to speed innovation, Verizon chose to locate in the tech capitals of America, attracting tech talent that might never have considered a phone company. Brian Higgins helped make that happen.

Higgins, who had flown to Boston from Verizon headquarters in Basking Ridge, New Jersey, to personally conduct a tour for me on a pleasant spring day in 2014, is vice president of mobile and internet services, a promotion from his job as executive director of LTE ecosystem development when I met him at the Innovation Center in 2011, two years after it opened, and just as CEO Ivan Seidenberg was passing the baton to new CEO Lowell McAdam, a veteran of the wireless side of Verizon. Higgins brought to fruition a wild idea hatched by two top Verizon technology leaders—an open center where any company in any industry could get free space and support to develop 4G LTE applications. He is proving that giants can learn to dance, doing cool things quickly.

Futuristic visions start in the Innovation Center's parking lot. Placed prominently at the end of the row closest to the entrance is GE Watt, a wireless network-connected electric vehicle charging station, in case visitors come in electric cars. GE Watt provides live usage data, enabling electric car drivers to use smartphones to find available charging stations.

Inside, Brian Higgins and two colleagues greeted me and a team member. The first thing we saw was a projection of the Harvard Business School logo on the floor by the entrance, a nice touch done for all external visitors with their own logos. Red LED lights throughout create a jazzy feel, and the clean modern lines make it easy to focus on the displays. Prominently in view was a connected vehicle, in this case a smart van. On a console nearby, I watched Network Fleet, an end-to-end vehicle fleet management system already in use by Verizon, municipal governments, and businesses. Like a few similar services running on the Verizon network, it shows live real-time data about the location of moving and parked vehicles, including fuel efficiency and speed. For pedestrians or bus riders, the center featured the HumanKiosk, a network-connected information booth with interactive displays. Perhaps bus stops can become community information centers.

Smarter trash was the next stop. The BigBelly solar trash barrel monitors capacity and, through the network, sends alerts to dispatchers. It takes a second for it to sink in that this is not just a clever gimmick. Smart trash barrels, especially powered by renewable energy, could reduce traffic by targeting garbage truck use. In fact, a major package delivery company saw the smart trash barrel on a tour of the Innovation Center, recognized the implications for its own drop boxes and truck use, and is now developing this technology with Verizon. Score two for reducing congestion and pollution.

The groovy stuff is fun. There are smart pill bottles that remind you when it's time to take a pill, or VGO telepresence robots that communicate remotely while mobile, which appeared in the movie *Iron Man 3*. These kinds of devices attract a now-constant stream of business visitors to envision new possibilities for use of mobile networks in their own industries.

But the real work takes place in about a dozen network-enabled labs in adjacent buildings off-limits for visitors, and sometimes so secret that they're off-limits for Verizon staff too. In those labs, developers from a wide variety of companies experiment with 4G LTE to invent new services and mobile applications. They enjoy access to Verizon's 4G LTE network test bed, private engineering spaces, and on-site Verizon network engineers and development staff. Some of the labs can run on slower 3G networks if needed, or European GSM networks, not just on U.S. systems. There is no requirement that developers use the Verizon network for their eventual applications. But by building the ecosystem for everyone, Verizon gets benefits too. Verizon can identify new products and more uses of mobile networks, which would, not incidentally, increase revenues. Some of the biggest hits are partnerships with other big technology companies like Cisco, Siemens, and GE for behind-the-scenes capabilities consumers can't see.

On my first visit in 2011, about one hundred partner companies had already been involved in various stages of exploration. Less than four years later, three hundred had been involved, Higgins said, with Verizon becoming choosier and more strategic. Since the start, seventy-five products have been commercialized, with ninety in the pipeline. In January 2010, then CEO Ivan Seidenberg appeared onstage for the first time at the Consumer Electronics Show in Las

Vegas with some of the coolest products, surprising those who were not expecting a once stodgy telecom to have widely appealing products to showcase. At the January 2013 show, Verizon awarded its first "Powerful Answers" no-strings grants to small companies targeting sustainability, health care, and education. There were 1,400 applications for the 2014 award. Transportation, including smarter driving and logistics, is a 2015 focus.

As Verizon helps build a mobile-based ecosystem of networks and applications, some collaborations are paying off for the public. Synovia Solutions partnered with Verizon Wireless to install more than 65,000 GPS units on school buses across North America. This enables parents to monitor their children's commute to and from school, and the public schools to ensure optimal use of buses and to reduce insurance costs. School districts with fleet sizes ranging from 25 to 3,000 buses are realizing annual cost savings from $50,000 to more than $1 million. Good for parents, good for the people.

In addition to supporting mobile apps, Verizon is working on machine-to-machine telematics. Telematics is an interdisciplinary field encompassing telecommunications, vehicular technologies, road transportation, road safety, electrical engineering (sensors, instrumentation, wireless communications), and computer science (multimedia, Internet). In 2012, Verizon acquired Hughes Telematics; its technology is the basis of the network-enabled fleet management systems I saw at the Innovation Center. The Verizon Telematics unit builds systems enabling car owners to turn off remotely a car that has been stolen or automatically call an emergency service in the case of an accident, competing with OnStar while innovating in a growing market. Because of its interest in supporting the connectivity of vehicles through 4G LTE and cloud-based services, Verizon is active in intelligent transport industry associations, such as ITS America and the Intelligent Car Coalition.

Verizon is not alone in this quest, nor is it unique in the industry—although what the industry consists of is not clear. Just as in car- and ride-sharing and parking services, there is growing convergence across industries, and there will be industry shakeouts. Just because Verizon and AT&T, the mobile network giants, emerged triumphant from generations of Ma Bell divestitures, telecom consolidations, and wireless transformations, that doesn't mean they will survive battles with cable companies over high-speed Internet, with IBM and Amazon over machine-to-machine connections through cloud computing, or with Google for nearly everything. Google, a force in mobile, has been laying fiber-optic cables to run faster data networks over landlines. This is classic "frenemies" territory, where today's partner becomes tomorrow's competitor. But it all adds up to the potential to manage mobility of the physical kind as a connected whole.

INTELLIGENT TRANSPORTATION SYSTEMS: GETTING ALL CONNECTIONS TO WORK TOGETHER

Network technologies have increasingly come to support the development of "intelligent transportation systems" across all modes of transportation, including passenger and cargo mobility, enhancing "the ability of physical transportation infrastructure through better management of flows and direction of traffic."[26] Rail, air, and public transit already have variants of centralized control centers or operations centers, for dispatch of vehicles and observing their interactions—air traffic control makes it possible for large numbers of objects to move through the skies, and towers at airports are in constant touch with operations centers such as the American Airlines facility I visited in Fort Worth. It seems only sensible to do the same thing with roads, as they come alive and are crowded with smart vehicles. With surfaces sensor equipped, and vehicles software guided and connected, shouldn't the roads have larger control centers to monitor and manage them? And shouldn't smart vehicles such as driverless cars have human vehicle traffic controllers in a tower somewhere, in case of technology malfunction?

Intelligent transportation systems pull everything together, using networks as the platform to facilitate the coherent interaction of the other three major components of the transportation ecosystem:

fixed infrastructure (roads, railway tracks, airport runways), vehicles (cars, trucks, trains, airplanes), and users (passengers, shippers).

An early example of an intelligent system is the tracking of surface shipments through use of GPS. UPS and Federal Express were
pioneers in the deployment of tracking technologies. By 1992, UPS
had integrated new technologies (creatively called handheld Delivery Information Acquisition Devices, or DIADs), electronic signature capture, and an electronic data communications network
such that it was able to track all ground packages. Federal Express
achieved similar technological advancements in the early 1990s
when it launched its MultiShip service, which was the first carrier-
supplied customer automation system to process packages shipped
by other transportation providers. Fully digitized ground tracking
made Federal Express, in 1994, the first shipping company to provide customers online tracking through its website. Technology also
permitted companies to monitor their drivers. Previously, truckers had significant autonomy to manage their own time, routes, and
speed. GPS fundamentally changed the playing field for the trucking
industry by enabling shippers to monitor truck locations, provide
drivers with more mapping and logistics information, or insisting
that they stop driving and get some sleep—the latter being especially
controversial among drivers.[27]

Early mobile technologies also set the foundation for connecting machines over mobile networks. For example, in 1996, Verizon
deployed 3G-enabled OnStar technology in General Motors vehicles
to enhance features such as mapping and emergency communications. OnStar now operates on AT&T's mobile network.

Governments around the world have been taking active steps
to promote intelligent transportation technologies, especially in
areas with dense populations and high demand for mass public
transit—European states, Japan, and China, which are also the
powerhouses in high-speed rail. Because these countries have a
stronger tradition of centralized government, implementation of
intelligent transportation systems across multiple transportation

modes, such as air-to-rail, has been faster there than in the United States, where the focus has been primarily on road traffic. In 1991, passage of the federal Intelligent Vehicle Highway Systems Act allocated an initial $250 million annually to research, pilot projects, and standard setting for intelligent transport systems. From 1994 to 2012, the U.S. Department of Transportation oversaw the allocation of more than $3 billion for intelligent transportation research and deployment. By 2010, state governments were spending an estimated $1.3 billion of their combined annual federal highway funds and another estimated $1.3 billion of funds from the American Recovery and Reinvestment Act on smart transportation R&D. As with NextGen technology for air transportation, road transportation provides another example of private-sector technology innovators and entrepreneurs racing ahead to create products while the public sector does research studies.

There are high hopes for the benefits. In 1998, European researchers estimated that, by 2017, intelligent transport would bring about a 15 percent increase in survival rates from car crashes, a 25 percent reduction in travel times, and a 25 percent reduction in freight costs. Although data are fragmented and often project or municipality based, many indicators show that the investment is paying off. A 2011 study by the Insurance Institute for Highway Safety in fourteen U.S. cities found that intelligent red-light enforcement programs using video cameras led to a 24 percent lower rate of fatal red-light-running crashes during a four-year period relative to baseline forecasts.[28]

There are also wider economic benefits from investments in system-wide technology. The U.S. Department of Transportation Research and Innovative Technology Administration estimated in 2011 that the intelligent transportation sector was already worth $48 billion annually and growing. The agency also estimated that the industry was providing 180,000 private-sector jobs and would continue adding between 3,600 to 6,400 new jobs annually through 2015.[29]

It's good to hear this kind of forecast, even if the numbers are soft.

Amid the growth of the Internet of Things, in which machines communicate autonomously with other machines without human intervention, it's heartening that robots won't steal all the jobs.

IS AMERICA READY?

Technology is an American strength. But while often the world's innovator, with top private-sector companies, the United States sometimes lags in deployment of innovations, such as widespread affordable broadband access. "We lead on innovation where we don't have coordination challenges," ITIF president Robert Atkinson says.

Broadband infrastructure, especially mobile broadband, is the backbone for new, innovation-rich, technology-enabled transportation systems to operate successfully and get to the quintuple wins. Demand continues to increase exponentially, as more users populate mobile networks and want higher speeds; analysts forecast that LTE will grow from 44 million worldwide users in 2012 to more than 1 billion by the end of 2017.[30]

Globally, America often ranks in the top ten among a variety of indices that measure technological progress. The highest U.S. rankings are those which give significant weight to technological innovation and industry-led R&D, including the Global Innovation Index, Global Technology Index, and IT Industry Competitiveness Index. With indices that give greater weight to nationwide network usage and deployment, the United States ranks relatively lower. On a widely cited United Nations International Telecommunication Union index of information and communication technology released in 2013, known as the IDI, the country ranks number 17 globally, behind Korea, Sweden, and Iceland in the top three, and Macau (China), Singapore, and New Zealand just ahead of the United States.[31] The IDI is a composite of three subindices. In the IDI "access" subindex, the United States ranks number 29, behind Barbados, Estonia, and Bahrain. The access subindex measures fixed-telephone subscriptions per 100 inhabitants, mobile-cellular telephone subscriptions per 100 inhabitants, international Internet bandwidth per Internet user, percentage of households with a computer, and percentage of households with Internet access at home.[32]

Whatever we have, it might not be enough. Jeffrey Immelt, chairman and CEO of General Electric, says, "Broadband in the United States is not where it should be." He should know. He leads one of the world's most prominent industrial giants with a major presence in transportation and infrastructure, and he chaired the President's Council on Jobs and Competitiveness in the years of recovery that followed the financial crisis. He also cites personal experience. As I mentioned in the first chapter, he recently had a weekend call to a prominent business leader in Nigeria to discuss a business deal in Africa, which he started in his car, but then had to tell the potential partner that he would call back from a landline phone at home because he had no cell access in his part of Fairfield County, Connecticut. Hard to believe. The Nigerian probably also found it hard to believe, undoubtedly having assumed that he was the one from the Third World country.

A cautionary note. The rankings could be out of date. Many ranking schemes reflect Great Recession data; advances in technology since 2007 (Apple's iPhone launch), cloud computing, and the like have helped America's broadband systems double in speed from 2010 to 2013. According to Information Technology and Innovation Foundation experts, America's broadband networks lead the world by many measures, and they are improving at a more rapid rate than networks in most developed countries.[33]

Verizon's McAdam feels stumped when asked why the United States often lags behind Europe in many technology rankings, especially because the country makes up over half of the 4G LTE mobile connections in the world. Noting the discrepancy between U.S. investment in broadband infrastructure and deployment, McAdam says, "For the past thirty years, we've been putting the tools in the toolbox; the next three to four years will need to be about how we leverage them." ITIF's Atkinson agrees that it's not our networks that are deficient; it's our application to transportation that lags. America's entrepreneurial spirit is soaring, but we can't count on entrepreneurs with parking apps to create the future. The future requires discussions of a sort that politicians seem to be ducking, about still-to-be-addressed business and policy challenges.

For any of this to work, it must work together. Our devices, apps, vehicles, and roads must be able to talk to one another, lest opportunities stop at each parking meter, toll road, or driverless car. Too many borders and boundaries limit opportunities. Of course, the same holds across modes, as trucks and rail intersect, and as passengers move from one form of transportation to another. Achieving full-scale interoperability, as it's called, across software systems, transportation modes, and geographic jurisdictions is a complex challenge. It can't be done by one company at a time, or they will each go their own way, and little will connect.

In other words, will rail play with roads, iPhones with Androids, GM with Ford, RideScout with ParkMe, Zipcar with Uber, California with Illinois? Not to mention whether the United States will play with the rest of the world, which uses a different set of mobile standards, known as GSM—global system for mobile? For transportation systems, which cross many jurisdictions and involve both public assets and private actions, legal standards and interoperability can't be left to each city council to determine—the way Uber, Zipcar, and other new smart transportation models are being handled today. It will require concerted national effort.

It takes more than a village. Isolated industry villages will have to turn into cross-sector multistakeholder coalitions pretty fast, building more collaborations like the Open Automotive Alliance. As more manufacturers seek to equip their products with network capabilities, the demand for gateways to access the wireless network will also increase. Verizon and providers of wireless services argue that a single gateway for multiple objects that are communicating with one another is the best solution for maximizing interoperability.

Such interoperability is achievable only when industry standards for wireless connectivity are set. Organizations such as the American National Standards Institute provide a neutral ground for businesses to collaborate on developing voluntary standards that will facilitate interoperability.[34]

In addition to standards, there are questions about capacity. Some experts question whether the Federal Communications Commission (FCC) has allocated sufficient wireless bandwidth to power widespread machine-to-machine usage, such as intelligent transportation systems. The greediest gobbler of the spectrum allocated for mobile use is streaming entertainment, such as music and movies. Transportation systems are considered less of a burden because of the relatively lower volumes of data transmitted. But as both grow exponentially, more spectrum will be needed to achieve the full potential of intelligent transportation and infrastructure. Thus, Verizon and other technology companies work closely with the FCC to evaluate which spectrum bandwidths are most conducive to powering which types of technology at high speeds and large capacity.

Some experts want the FCC to allocate more unlicensed spectrum for public services such as transportation, ensuring lower cost than broadband services offered by private-sector companies.[35] But if that drives traffic to those bandwidths, it could clog the network, causing virtual traffic jams. Congestion on high-speed networks when large amounts of data are being sent can cause outages. In the first half of 2013, Verizon was reportedly caught off guard by the onslaught of data traffic (largely from videos) on its network and had to roll out additional spectrum in fifty U.S. cities.[36] Federal agencies such as the FCC and USDOT plus technology and transportation industries must decide collaboratively the best direction for spectrum policy to support intelligent transportation systems, which will certainly include NextGen for air transportation, as I described in the preceding chapter. For example, deploying technology to replace or augment the waning gas tax with vehicle-miles-traveled fees could ensure funds for repairs and upgrades, enable congestion pricing, and attract private investors because of the clear revenue stream from user fees.

Technology always moves faster than our ability to set standards for it. In mid-2012, young lawyers were vying for rules-writing jobs in the U.S. Department of Transportation, figuring out standards for vehicle communications, smart roads, delivery drones, and driverless cars. By communicating with one another, vehicles can automatically prevent crashes and reduce the risk of death or injury. But without standards, it's hard to know whether they might also *cause* crashes, death, or injury. The National Transportation Safety Board wants cars, trucks, and buses to be preequipped with vehicle-to-vehicle communication technology and has recommended that its counterpart, the National Highway Traffic Safety Administration, develop standards for connected-vehicle technology.[37] Do we want Amazon's delivery drones dumping boxes of books or diapers in our neighborhoods, risking the heads of the children who will be read to or diapered? And all that data in Big Data is another issue. Smart transportation pinpoints the physical location of users, through the apps, vehicles, and sensors. Where will the information from sensors go, and who will have access to it?

I imagine that most Uber users are too busy getting where they need to go quickly to stop to notice sensor-equipped roads, see Google car test drives in their area, or think about these kinds of questions. But without answers, we will still be stuck en route to the future.

MOVING ON MOBILITY: A NATIONAL PRIORITY

Moving people and goods is now inextricably intertwined with moving information. The Information Superhighway directs the highways. Mobile communications systems are reinventing transportation systems through smart systems for roads, airways, railroads, and vehicles, promising solutions to traffic congestion, fuel efficiency and pollution, productivity losses caused by delays, and other pain points and bottlenecks. This is a domain with abundant opportunities: for business innovation and growth, as new business models follow new technologies; for communities and regions to improve quality of life, as they help people become more mobile with greater convenience and safety and fewer environmental hazards; and for national economic growth and competitiveness in the future. Verizon's McAdam says that if the United States decides to improve intelligent transportation, technology companies will rally, and it will get done. He also promises that if NASA wants cell sites on Mars, Verizon will build them.

U.S. Senator Edward Markey recalls how America leaped ahead with widespread development of the Internet. As a congressman, he was active in rewriting communications laws to open access to the Internet, believing that competition would stimulate innovation and benefit consumers, as long as rules were clear and companies met their obligation to contribute to education and make access affordable for schools. He says that the same thing should happen with transportation now. Let government create the rules, and then get out of the way to allow innovation to flourish.

But in America, the rules of the road are slow to catch up to the potential. ITIF president Atkinson complains that only 2 percent of transportation-related spending in the United States is used for intelligent transportation. In contrast, Singapore has also been very active in deploying V2V and autonomous vehicle technology to improve taxi and public transportation. The government has prioritized transportation as a national policy issue and allocated significant funds to attract American universities to lead the way, including MIT. MIT's Computer Science and Artificial Intelligence Lab chair, Professor Daniela Rus, reports that with access to a data set of 26,000 Singaporean taxis, they can potentially improve efficiency and reduce the number of taxis on the road by 70 percent.

The intersection of transportation and communication should be a national discussion across the private and the public sectors, start-ups and established players. It cannot be left to federal regulators and the courts, nor to various sectors operating in isolation, one issue at a time. Technology for transportation should become a priority, targeted in every digital or innovation initiative, including Small Business Innovation Research grants and Small Business Administration loans. There should be joint U.S. Department of Transportation, Department of Commerce, and FCC initiatives. The Highway Trust Fund should be renamed the Mobility Fund and prioritize strategic projects that involve technology. Innovations by private entrepreneurs can come to scale only if the public sector invests in facilitating platforms and connection points.

Although moving people and goods won't happen without information infrastructure, these interdependent industries rarely communicate. America needs a cross-sector, cross-industry discussion involving entrepreneurs, large companies, and visionary public leaders to identify priorities and advocate for change. Although I'm reluctant to call for another national commission that goes nowhere, at least the members could arrive in shared cars that find their own parking.

5

RETHINKING CITIES

Transportation dominates daily routines. It opens opportunities or constrains choices. And its possibilities shape cities. Cities are transportation hubs, centers of commercial exchange, and the locus of lives. All roads lead to cities; all modes meet in cities. From ancient times, concentrations of population have grown around transportation intersections—crossroads and waterways, and, later, train stations and transit stops.

The problem is that ancient times still prevail in many American cities. The basic designs of rails, roads, bridges, and city streets were laid out a century or more ago. Sixty years ago, large investments in highways took people away from cities and increased the growth of car-dependent suburbs. Congestion, pollution, and delays grew, while public transit systems fell behind, serving fewer people, and not always those who needed it the most. Transportation modes developed haphazardly, one isolated segment at a time, not connecting with other modes and sometimes interfering with them. Multimodal people were living in a one-at-a-time world.

Since then, technology, urban patterns, and cultural expectations have changed. A shift is under way from the centrality of cars, a quintessentially private mode, to a reemphasis on public transportation and people-friendly roads. Car-centric cities are no longer sustainable, either financially, environmentally, or socially. Cars and trucks won't disappear, but they must move over to respect other modes of getting around in cities.

But when infrastructure is also ancient, it's harder to think about the future. In 2012, a quarter of all U.S. bridges were deemed structurally deficient or functionally obsolete. Buses typically have a useful life of a dozen years but are still in service after twenty-two or more years. In Chicago, some railcars are thirty years old or older, well beyond reasonable use. The bulk of railroad infrastructure is well over a hundred years old, matching the age of some remaining wooden sewer pipes made from hollowed tree trunks. The consequences are like a horror movie: *The Street That Devours SUVs*. Chicago's ten-year plan to replace 900 miles of water pipes and repair 750 miles of sewer lines is addressing the joint challenges of high rates of water loss (490 million gallons a week), decreased water quality, and deterioration of streets, in one instance creating a gaping hole that swallowed a sports utility vehicle.[1] Different dangers lurk on city streets for unlucky people who are the victims of vehicle accidents.

Cities are changing in glorious ways, as hotbeds of creativity alive with opportunities 24/7. But infrastructure and transportation policies are only slowly catching up to the magnitude of the changes. Repairs alone will not be enough to meet current and future needs. Not only is infrastructure outdated; so are many cultural assumptions. It will take a dramatic rethinking of how streets are used and how people move. *Old thinking*: how to get people out of cities. *New thinking*: how to keep people in cities and help them move around. *Old thinking:* streets are for cars. *New thinking*: streets are for people. These seemingly simple ideas have profound implications for investments and leadership for the future of American cities.

With new ways of thinking about them, cities can achieve the quintuple wins project by project: greater safety, less congestion, higher productivity/efficiency, less pollution/carbon emissions, and more economic opportunity. They can be bigger in potential as well as in population.

BLIGHT TO DELIGHT: THE CULTURAL
CONCEPT OF CITIES

Over the past sixty years, the proportion of the world's population living in cities and their surrounding metropolitan regions has grown dramatically, from a third in 1960 to over half in 2014; the urban population is projected by the UN to grow to two-thirds by 2050. Urbanism figures are much higher for the United States: 70 percent of the population already in cities in 1960, growing to 81.6 percent in 2014, with projected growth to 87.4 percent in 2050.[2] The U.S. definition of urban is at least 2,500 inhabitants with density greater than 1,000 per square mile, and it includes small towns as well as suburbs ringing the central city.

American central cities are making a comeback after decades of suburban growth. That trend offers challenges and opportunities for transportation. Since the 1920s, U.S. suburbs have grown faster than city centers in every decade; in 1970, the ratio of U.S. central-city populations to their total metropolitan area populations was 0.43. In 1980, the ratio fell to 0.38. The ratio continued to drop, albeit more slowly, to 0.34 in 1990, 0.33 in 2000, and 0.31 in 2010.[3] Suburbia has been the residential location of choice. But in 2010, the situation started to flip; for the first time since the 1920s, population growth in the urban cores has outpaced growth in the suburbs. Since 2010, big cities have gained more people than during all of the preceding decade. Taken together, cities with populations greater than 250,000 grew more than 1 percent annually since 2010—double the preceding decade's growth rate. National averages would be higher if not for the abandonment of financially distressed cities such as Detroit.

Residential patterns are bringing the more affluent back to cities, including young adults and young families, for whom suburbia is no longer the only choice. Part of what's happening is groups changing places, with the less affluent retreating to aging inner suburbs while downtown and adjacent areas gentrify. In metro areas with population over one million, suburban growth has hovered at about

0.95 percent annually, compared with a 1.38 percent annual average since 2010. The fastest-growing city centers relative to their suburbs are New Orleans, Charlotte, Seattle, Minneapolis, Columbus, Richmond, Washington, San Diego, Denver, and Boston. New York and Philadelphia have also seen small gains.[4]

"Chicago has the most rapidly growing population of any downtown area in the country," declares Deputy Mayor Stephen Koch, perhaps exaggerating in his enthusiasm, with a different set of numbers in mind. "It's largely millennials, people between the ages of twenty-five to thirty-five. They want to live and work in close proximity. They want to commute in a much more sustainable way." The cultural shift is a greater truth than the numbers imply. It suggests a new image for downtowns as places to both live and work.

THE ATTRACTIONS OF CITIES

Suburbia was utopia in the 1950s and beyond, a placid place with green space and good public schools. But cities were viewed as sources of sin and blight. The word "urban" was almost inevitably followed by the word "problems." In the 1950s and 1960s, urban renewal consisted of razing working-class neighborhoods or older parts of cities to erect sterile concrete structures, whether highway overpasses, public housing, luxury high-rises, or empty plazas. Cities saw in-migration by immigrants and people of color, along with white flight (or flight of the middle class) to bedroom suburbs. Workplaces were separated from families and schools; long commutes to work became the norm. Health care was concentrated in large hospitals and their emergency rooms, which were centers of care for urban populations but at high cost, as ER care is expensive. And transportation policies were often designed to help people escape center cities, leaving streets empty and dangerous at night.

Starting in the 1950s, national and state policies favored cars and roads, within cities as well as outside city limits. Transit users and pedestrians were distant afterthoughts, and bicycles a suburban toy. Car ownership was growing along with bedroom suburbs, reinforcing an assumption that it should be easy for people to take roads, especially the new superhighways, out of cities. Suburban commuters would come to work downtown, park, and leave again using the highway systems. As road use became easier, more people drove, which exacerbated a growing congestion problem, making traffic tie-ups and delays seem normal, and creating smog and pollution. Then, because streets were considered habitats for cars, they did not always work well for people. It's risky for pedestrians and cyclists to use city streets.

The role of cars and the institutions they've formed is diminishing. Every year since 2004 has seen steadily less car ownership and less driving, making it clear to leaders in many cities that streets should accommodate users other than cars. That matches a quiet revolution in values and a changing concept of what cities are and should be.

Cities are increasingly viewed as sources of innovation and quality of life. There has occurred a rediscovery of the role of density and diversity in creativity.[5] In contrast with the industrial economy or mass production economy, cities are centers of the service economy, experience economy, sharing economy, entrepreneurial economy, and personalized economy. Pick your favorite name; they all suggest an image of the future in which people are out on the streets. New-technology industries that employ young talent favor urban workplaces, causing cities to name innovation districts where entrepreneurs and innovators bump into one another on the streets and in restaurants, bars, and coffeehouses. Like individuals, institutions increasingly intersect. Businesses and other employers are involved with urban schools, sending employees to tutor and take schoolchildren to workplaces through partnerships with nonprofit organizations offering mentoring opportunities.

The image of cities is changing from being centers of blight to being places with experiences that delight. Cities are increasingly expected to be people friendly, environmentally friendly cultural centers, in which neighborhoods participate in "placemaking," a creative enrichment of small areas, often former parking spaces. There is a greening of cities through miniparks, rooftop gardens, and use of vacant lots for community gardens; a desire for safer streets that are pedestrian and bicycle friendly; and a small but growing ethos of sharing rather than owning. Cities are flourishing by becoming arts centers, attracting creative talent and tourist dollars, and hosting street festivals whenever the weather permits. Former warehouse districts are becoming homes to artists, galleries, museums, music schools, and performing arts centers; Miami, for example, has reinvented itself using an arts-based strategy that includes both physical infrastructure and global events, such as Art Basel Miami. From San Francisco's Mission Street renewal to Boston's downtown Rose Fitzgerald Kennedy Greenway, a ribbon of imaginative parks made possible by the Big Dig, cities are becoming more attractive, vibrant, and livable.

Once viewed as sources of environmental problems, cities are increasingly seen as solutions. Douglas Foy, former CEO of the Conservation Law Foundation, environmental chief for Massachusetts, and architect of the green plans for New York and other cities, is one of the environmental experts who believe that the density of cities can help reduce energy use and carbon emissions, as power sources are shared and reliance on cars is reduced. People can more easily take care of one another, Foy says, feeling that cities are the answer to almost any question. Urban economist and Harvard colleague Edward Glaeser has called cities the healthiest, greenest, and richest places to live, providing the best hope for the future.[6] Glaeser also suggests that instead of building out—suburban sprawl—cities should build up—increasing density per square foot of ground. In his view, elevators should be recognized as a mode of transportation. He's serious. Elevators can carry people from homes to workplaces in high-rise complexes or among segments of vertical factories.

Ah, the glories of cities. More accurately, the potential glories. Many solutions are still works in progress. Cities remain riddled with problems. They are dependent on state and federal aid. They face the politics of suburbs versus cities—suburban commuters who benefit from the urban core don't want to pay for it. Urban neighbor-hoods are marked by group versus group tribalism. The most pros-perous cities suffer from inequality and inequity. The income gap is widening, and the benefits of new city amenities go to some social categories more than to others. But cities are also embracing new possibilities with potentially big payoffs.

There are two big opportunities for American cities of the future: upgrading and reinventing public transit; and reinventing streets to be people friendly. To act on those opportunities requires many actors: top leaders, community activists, entrepreneurs with new business models, politicians, planners, civic groups, and ordinary people. Models from other countries can be a guide to help American cities transform their cultures and structures, and create a platform for leaping ahead.

THE CASE FOR PUBLIC TRANSIT

The "public" in public transportation covers a wide swath. More than 7,200 organizations provide public transportation in the United States, in the aggregate employing more than 400,000 people in a $57 billion industry, according to the American Public Trans-portation Association. But in the aggregate this is still less than one major Fortune 10 corporation. The case for public transit is not that it makes money from operating the service; it generally doesn't, and even earn-as-you-operate services can be highly subsidized by tax-payers. What matters are the other benefits.

Public transit is worth a great deal to a city. Hidden economic value can range from as much as $1.5 million a year for the smallest of cit-ies to a whopping $1.8 billion a year for the largest cities, according to urban economists Daniel Chatman and Robert Nolan, who analyzed 2003–07 data from 290 metropolitan areas. Their study shows that adding about four seats to rail lines and buses per 1,000 residents produces 320 more employees per square mile for the central city, an increase of 19 percent. Adding 85 rail miles delivers a 7 percent increase. A 10 percent expansion in transit service by adding rail and bus seats or rail miles produces a wage increase between $53 and $194 per worker per year in the city center, and an associated increase of 1–2 percent in the gross metropolitan product, a measure of urban economies equivalent to the gross national product for countries.[7] These figures predate the financial crisis and Great Recession, of course, but the sheer magnitude of the effects is impressive.

It's well known that economies grow around train stations and airports, and that property values increase in areas adjacent to good transit connections. The American Public Transit Association claims that every $1 invested in public transportation generates approximately $4 in economic returns. Their figures show that every $10 million in capital investment in public transportation yields $30 million in increased business sales, and every $10 million in operating investment yields slightly more, some $32 million in sales increases. From 2006 to 2011, residential property values performed 42 percent better on average if they were located near public transportation with high-frequency service. There are also cost savings: in 2011, U.S. public transportation use saved 865 million hours in travel time and 450 million gallons of fuel in 498 urban areas. Without public transportation, congestion costs in 2011 would have risen by nearly $21 billion, from $121 billion to $142 billion, in the 498 urban areas.

Furthermore, demand is increasing. In 2013, Americans took 10.7 billion trips on public transportation, the highest in fifty-seven years, and 35 million people a day boarded public transit. Since 1995, public transit ridership is up 37.2 percent, outpacing population growth, which is up 20.3% percent, and outpacing auto use growth measured by vehicle miles traveled, which is up 22.7%, according to American Public Transportation figures. The call for more and better public transit is surprisingly universal. It's first on the transportation and infrastructure wish list for a wide range of business leaders. Nearly 40 percent of the respondents to a Harvard Business

School U.S. competitiveness survey in December 2013 selected this as a top three priority out of a dozen possibilities, getting the highest percentage of any item. This concern is largely on behalf of their employees and how they get to work. Public transit, also known as mass transit, is an urban necessity.

HOW SUBWAYS AND BUSES GOT TOGETHER:
SOME RAPID HISTORY OF RAPID TRANSIT

On September 1, 1897, in an effort to ease streetcar congestion, a private company opened the Tremont Street Subway in Boston, later named the Green Line, after the Emerald Necklace network of parks that it bordered. This wasn't the world's first tunnel for trains; London created its Underground decades earlier. But it was a national first. New York followed with its first underground line in 1904,[8] and the rest is history—single lines became multiple lines crisscrossing both cities. Later, other American cities built subways or heavy-gauge elevated railways, such as BART (Bay Area Rapid Transit) in San Francisco. Metrorail, in Washington, D.C., completed the last segment of its original planned line in 2001, the most recent full subway system to be built, and undertook extensions to be completed in 2014 and 2016.

Subways are expensive to build and risky to operate, as the fate of the Green Line and her sisters illustrates. In 1947, because of increasing automobile usage, the privately owned subway was in financial distress. The Commonwealth of Massachusetts acquired a majority stake and formed a regional public transit authority to run it, the Massachusetts Transit Authority, featured in the Kingston Trio song "Charlie on the MTA." The MTA that the fictional Charlie rode forever 'neath the streets of Boston became today's Massachusetts Bay Transportation Authority in 1964. The enlarged name reflected additional acquisitions to ensure connections among subways, suburban buses, and commuter rail over a wider geographic span. More MBTA takeovers followed. During transit-oriented Michael Dukakis's first term as Massachusetts governor (he proudly rode the Green

Line to work) in the late 1970s, he hired former CIA agent and Boston deputy mayor Robert Kiley as MBTA chief to fix the system (Kiley later ran the New York City subway and London Underground). With the backing of then state transportation secretary Fred Salvucci, the MBTA bought the commuter rail system for the bargain price of $60 million. The MBTA has required an annual operating subsidy from the state legislature for nearly everything—although in 2014, as Richard Davey, the current Massachusetts transportation secretary, proudly told me, the MBTA was able to pay its employees from its operating budget for the first time in twenty-four years.

Regional authorities are now the norm for owning and governing subway and bus systems, even when private contractors take on portions of their operation. Operating losses require subsidies, and capital comes from the public side. New projects appear massively expensive now, even when they build on existing systems. In San Francisco, the 1.7-mile BART Central Subway Project is estimated to cost $1.7 billion—or $1 billion per mile. Washington Metro's Silver Line extension to Dulles Airport, a 12-mile above-ground project, will cost $5.6 billion, while a 2-mile section of New York's Second Avenue line, currently under construction, will cost $4.5 billion— more than $2 billion per mile. One study found that subway projects in the U.S. cost three to four times as much as comparable projects abroad. And then there's the fact that, once built, subways are fixed in place. If you don't build it in the right spot, you risk ending up with a megadollar system that practically nobody uses, like Detroit's 2.9-mile People Mover monorail, built in 1987 at a cost of more than $425 million in present dollars.[9]

Massachusetts has played a disproportionate role in the financing of public transit too. In 1973, thanks to the work of its congressman Thomas P. ("Tip") O'Neill, then majority leader of the House of Representatives, with active bipartisan support from its Republican governor, Francis Sargent, the Federal-Aid Highway Act allowed states for the first time to divert Highway Trust Fund monies from highways to transit projects. During conference negotiations over the Federal-Aid Highway Act of 1972, O'Neill worked to ensure that a Senate bill for transit subsidies would become law; when it grew clear that his fellow conference committee members would refuse to compromise, he helped to scuttle the bill and declared that the HTF was "no longer a sacred cow in Congress." That maneuver worked, and his preferred bill eventually passed the next year.[10]

A decade later, with O'Neill elevated to Speaker of the House, the 1982 highway bill created an official transit account, which receives about 16 percent of Highway Trust Fund money. Many people credit Salvucci, with Dukakis's backing, for active behind-the-scenes work in support of the bill. Supplemented with $5 billion general fund transfers in fiscal years 2013 and 2014, the transit account allocates about $10 billion annually to mass transit projects across the country, primarily through the Federal Transit Administration. In 2013–14, notable projects with more than $100 million in funding through the transit account include Denver's Eagle P3 commuter rail project, the Long Island Rail Road East Side Access project, the Honolulu High Capacity Transit Corridor Project, the Silicon Valley Berryessa Extension Project, San Francisco's Third Street Light Rail Central Subway Project, and Seattle's University Link Light Rail Extension, as well as extensions to the MBTA and renewal of historic inner-city stations.

IS PUBLIC TRANSIT A RIDE OUT OF POVERTY?
PHYSICAL MOBILITY AND SOCIAL MOBILITY

The social consequences of public transit for city residents might be as important as the economic consequences to city coffers. Opportunity for access to the tools of upward social mobility—jobs, a good education, health care, affordable groceries—is activated by geographic mobility. Inequality in America is a big problem, and there are arguments that the poor have remained stuck for decades. One solution might be to improve public transit. It turns out that half of the top twenty cities for intergenerational social and economic mobility are also half of the top twenty cities for public transit.

Where do children have the best chances of moving from poverty to affluence? Harvard economics professor Raj Chetty has examined intergenerational mobility by geography. His team studied all children born in the United States between 1980 and 1982 and grouped them by location at age sixteen, sorted into 741 "commuting zones," similar to metropolitan regions. Those children's income at age thirty in 2011–12 was compared with their parents' status.[11] This provided a ranking of the cities whose citizens have the best chances of making the big leap from the bottom fifth to the top fifth of the income distribution.

And where is there the best public transportation? A rating company, Walk Score, took data from 316 U.S. cities and thousands of neighborhoods and assigned a "Transit Score" to each, out of a possible 100 points. The Transit Scores consider frequency, type of route, and distance between stops.[12] Note that no U.S. city scored above 90, making it a Rider's Paradise of world-class public transportation, in the raters' lingo, but a few scored above 80 for Excellent Transit, convenient for most trips.

If we cross-reference these studies, the overlap between opportunity and public transit is striking: five of the top ten cities for physical mobility are also in the top ten for social mobility—New York, San Francisco, Boston, Washington, D.C., and Seattle. Close behind them are four other California cities—San Diego, Los Angeles, San Jose, and Sacramento—and Portland, Oregon, which are in the top twenty on both lists. Looking at this another way, we find that seven of the top twelve cities for upward social mobility are in the top twenty for transit scores. Seven of the top ten for transit are in the top twenty for upward social mobility.

There's a clear tech-economy effect. San Jose, California, in the heart of Silicon Valley, has the highest percentage of children who grew up in the bottom fifth reaching the top fifth, at 12.9 percent. It's nineteenth best for public transit. Neighboring San Francisco came close, with 12.2 percent of those starting out at the bottom reaching the top, and gets the second-highest public transportation score at

80.5, close to top-ranked New York at 81.2. Third-place Boston is also a center for technology plus higher education.

This overlap of opportunity and public transit is a startlingly high correlation. Of course, it's just suggestive, and many other factors play a role. Philadelphia and Chicago rank fifth and sixth nationally in transit but don't appear in the top twenty places for social mobility. In Chicago, there is a great gap between public transit's availability in general and its service to less-advantaged areas. The current transit system in Chicago fails to efficiently connect to employment opportunities outside of the downtown area. Chicago ranks fifty-third among U.S. metropolitan areas in labor market access, with only 22.8 percent of residents able to reach the average job using public transit in ninety minutes or less. In 2012, Chicago had the third-highest unemployment rate in the United States at 9.8 percent, down from 11.5 percent the preceding year. Chicago's most impoverished neighborhoods on the South and West Sides are predominately black and Latino and have areas where 50 percent of the adult population has been unemployed for over a year. The average black worker earns only 45 percent of what the average white worker makes, and all indicators of inequality show very little change since the 1960s. Neighborhoods on the verge of gentrification in 1995 continued their upgrade only if they were at least 35 percent white; those 40 percent or more black remained stuck, as studies by Harvard sociologist Robert Sampson show.[13]

Some large urban population segments are stuck in areas with high rates of violence, facing long commutes to jobs with little public transportation help. The evidence is consistent. A survey of Latinos in low-income areas in Massachusetts found that limited quality public transportation adversely affected their finances, job choices, and ability to get to health care appointments.[14] It's a sobering pattern in numerous cities. "Many low-income people and communities of color have had their transit services cut. Low vehicle ownership rates and lack of proximity to employment, daycare facilities, and grocery stores in these communities present a real problem," reports Anita Hairston, cochair of the national Transportation Equity Caucus.

But when it's possible to move around by public transportation, horizons widen and aspirations soar. Just ask Martin Johnson (not his real name), an intelligent, motivated black teenager living with a single mother in a remote, run-down part of Boston. He feels that public transit can propel him out of poverty to professional jobs.

Every day, twice a day, Martin rides two buses and a subway, walking at each end of his trip, to attend a school preparing him for an even better high school. It takes him nearly an hour and a half to travel six miles—if the buses are on time. It would be faster to walk if not for his heavy backpack and some dangerous highways on the way. For him, public transportation is a necessity for school; and beyond school, it provides access to a different kind of education, about life possibilities.

"I like having a glimpse of what I'll have in the future," he says. "If you take a really long bus, you can go from somewhere where it's not a good neighborhood to somewhere totally different. Sometimes I take the 66, which starts off at Dudley Station, which is the poor part of Boston, and it takes you all the way to Brookline (an affluent suburb). And the changes are drastic. There are times when I'm going home and I get on the 32, and there are kids rolling up marijuana. And I'm sitting between them and feel uncomfortable. That's not the environment I want to be in. I like the 47 bus which has doctors, people with suits and ties."

How many of us resonate with the idea of the bus as a tool for self-education? I know I do. As a middle-class preteen and teenager in a midwestern suburb, I found freedom and opportunity on the bus and rapid transit to the cultural richness of the city, from art museums to ball games to part-time jobs downtown. Unfortunately, cities have let transit systems slip.

THE GOOD, BAD, AND UGLY OF USING PUBLIC TRANSIT

In six cities—Boston, Chicago, Detroit, Los Angeles, Miami, and Philadelphia—my research associate Ai-Ling Malone and I met dozens of people, from teenagers to workers, who live in low-income areas and are not as fortunate as Martin Johnson. These commuters and students devise routines so that they can put up with making at least two and as many as four connections, plus time on foot, and they plan their days accordingly. Ninth graders do this for the chance to attend better schools. Workers do it for jobs with decent pay, which don't exist close to home.

People with limited income have limited options. Sometimes they walk. We heard from some who walk five miles to get downtown from their homes, because it's more direct and faster than multiple subway or bus transfers. They can't afford cars. For them, lack of car ownership is an economic necessity, not an environmental statement, which it might be for more-affluent young professionals. If it costs $800 a month for a car, then an $80-a-month transit pass looks awfully good. Not to mention the danger of getting tickets if Driving While Black, which risks higher insurance costs and license suspensions. So public transit claims them, inconvenient as it is. Some still long for cars. A worker in Detroit, when asked about his transit improvement wish, answered "getting a car."

In the best systems (or on good days anywhere), trains and buses are fairly clean and run roughly on time, with courteous drivers. That's mostly a fantasy. In the worst systems, ones that have been deteriorating and not supplemented by new services, trains and buses are old, dirty, and unreliable. Buses designed to run twelve years are still in service twenty years later in many U.S. cities. Neglected maintenance leads to breakdowns, which throw everything off, even if a backup is sent. This seems to happen not just in problem-ridden cities such as Detroit but in some neighborhoods of cities known for otherwise great transit, such as the South Side of Chicago or the minority-populated parts of Boston, leading to suspicion that they've been allowed to deteriorate because of racial prejudice.

The worst systems have surly drivers who sometimes don't finish their routes, because the area can be dangerous late at night and no one is riding, but they get away with it because of a code of silence.

Drivers disobey rules, such as those banning cell phones. "You've got some that just don't care," lamented a sixteen-year-old female high school student. "They'll pass your stop up just to stay on schedule or if it's too crowded. And people running for the bus, they keep going. I normally don't miss the bus, because people step out in front of it and make it stop."

But drivers, too, are sometimes the victims of an anticulture of disrespect and retaliation. We heard about a bus driver who had bleach thrown in his face when he wouldn't let someone on the bus. We heard tales of bickering and hostility, sometimes leading to violent attacks. Fights could bring in the police, which caused further delays and system backup because drivers were required to stay with the bus and file reports. For passengers, overcrowded conditions made one person angrily call his driver a number of ugly racist names. Another said it is the "wild, wild west" at some bus stops early in the morning when it is still dark, and thefts abound. Perhaps our informants were dramatizing for us, but these tales echoed across cities. Whether dangers were real or imagined, parents passed on to their children concerns about strangers: old men, maybe homeless, eyeing young girls, or bearing bedbugs or hidden germs.

There is little room to carry anything on subways and buses, although that's getting better. Some bus lines have added bike racks, but not enough. If there are only two on the bus, and they're used, cyclists must wait for another bus. In some cities, buses and subways have added racks on which to place bags, but many people say they still can't use mass transit for shopping, because it's not safe to wait with bags, there's no room for bags, and the whole trip takes much longer than by car. Impatient drivers and passengers don't always want to wait for moms to break down a stroller to bring it on board.

It's not just the ride that bothers riders; it's the wait. "People get frustrated," a twenty-six-year-old employed man said. "You're standing outside, and its a five-degree day, and you're waiting on a bus that's supposed to come every fifteen minutes, and one doesn't come for an hour. Those people get on, and they get upset. They get into arguments, get into fights." Passengers often don't know how long the wait will be, even if they are digitally enabled and have checked online for the latest information, which they've learned isn't always accurate. Train and subway stops are sheltered. Bus stops might not be. They lack roofs, benches, or places to sit (let alone the connected kiosks some tech mavens imagine for cities). In bad weather, if people seek shelter in a nearby doorway or gas station (assuming they're not chased away), they might miss the bus. Buses might slow down but not stop. One person said she would have a very long wait in cold rain or snow only to see buses zipping by in the other direction. A teenage student sometimes coped by boarding a bus in the wrong direction so that he could reach a sheltered transfer point.

Public transit not only takes people to work; it often serves as a workplace. Commuters check the news and get ready for meetings. Teachers finish grading papers. Schoolchildren do their homework— or, as some schools call it, their trainwork. And people adapt because they have no choice.

Community advocates believe that there should be a choice: get organized, and demand better. It's the American way: have an issue, form a group. The national Transportation Equity Caucus in Washington, D.C., has over seventy organizations as members, many of them large and influential, with substantial memberships of their own. These groups collectively advocate for public transit improvements to reduce disparities, and for streets that accommodate all users. There are also hundreds of grassroots, community-based advocacy organizations across the United States.

For example, millennial Miami activist Marta Viciedo leads a grassroots group called the Transit Action Committee, or TrAC. TrAC organizes colorful transportation awareness events and lobbies elected officials to support public transportation along with bicycles and pedestrians—Viciedo gave up her own car to prove the virtues of relying on public transit. A TrAC gathering might send groups across wide streets, to educate drivers about pedestrian rights (my name for this: Occupy Crosswalk), or commit Random

Acts of Transit (their name), such as beautifying downtown bus stops. Viciedo wants funds intended for public transportation to get there; she told us that a half-penny county surtax created for transportation was used to fund roadways until TrAC succeeded in persuading officials to make the necessary changes.

Slowly, transportation advocacy groups propel change. A decade ago, City of Miami mayor Manny Diaz couldn't get agreement from the City of Miami Beach for a light rail system between the two jurisdictions. In 2014, after a decade of unnecessarily awful road congestion on the causeways between them, it could finally happen. That could also invite children in Miami's impoverished inner city to see the ocean, just a few miles away, some for the first time. Maybe their horizons and aspirations will enlarge, like those of transit-riding Martin Johnson in Boston.

WHY FIXING PUBLIC TRANSIT IS DIFFICULT: POLITICS-AS-USUAL OR COLLABORATION

Beverly Scott has seen public transportation on the ground from leadership positions in transit authorities in California, D.C., Georgia, New Jersey, New York, Massachusetts, Rhode Island, and Texas. When she talks transit, people listen. The lesson from her experience is that public transportation can't be fixed when there's politics as usual, and that's how America has gotten stuck. Collaboration across sectors and jurisdictions is essential.

When Scott entered a doctoral program in government and public affairs, she had no intention of going into public transit. But she didn't want to be in human resources or public relations either—the typical roles for women or people of color in the 1970s. She identified transportation and sanitation as topics for a fellowship project. Transportation won.

Scott got her feet wet during eighteen months in Houston. The city's focus was on building a broad coalition to pass a referendum to create a transportation fund using 1 percent of sales taxes. It needed two rounds of votes. It passed the first vote, after a second attempt, because it was part of a broad referendum, but the experience gave Scott her first learning about the ups and downs of cities. The proposed project was called "the Saudi Arabia transit" because of its easy riches (a half billion dollars in the bank, some of it spent on a seven-floor building). About 20 percent of the contracts were designated for women and minority-owned businesses, a positive element "unheard of at that time," Scott recalls. The glitch: it lost the second vote, on the specific transit plan and funding, and the city didn't build a light rail transit system for twenty years. Why? There was growing opposition from the inner city. The plan was unconnected to the grass roots and contained elements, such as park-and-ride lots, that urban commuters didn't care about. Decades were lost because of a failure to fully engage the hearts and minds of the community. Scott carried this lesson with her to later jobs for transit authorities in New York, New Jersey, Washington, D.C., Rhode Island, Sacramento, and eventually the plum positions as CEO/general manager of Atlanta's MARTA for five years and then, since 2012, of the MBTA, based in Boston.

In 1996, when she started as general manager of Rhode Island's Public Transit Authority (RIPTA), President Clinton was overseeing the passage of a welfare-to-work bill to move people off welfare and back to work. But of course, they first had to be able to get *to* work. Rodney Slater, who was appointed secretary of transportation in Clinton's cabinet, still talks about his frequent messages to his department that transportation was the "to" in welfare-to-work. In Rhode Island, the welfare reform planning committee knew that transportation was among the top three issues, but even so, RIPTA was not at the table initially—only later, as the realities of implementation sank in. Dr. Scott was stymied by silos.

Later, in Atlanta, a different set of boundary challenges stemmed from the cross-county nature of the public transit plan, which linked the urban core and rapidly developing exurban and rural areas. Race and class issues were involved, because the old freight lines on which it would run were the dividing lines between communities, as were intercommunity equity concerns. Although MARTA started with a five-county vision, and four of five counties funded the original engineering report, two counties didn't contribute to the next phases, even while their board representatives continued to vote. The other two counties, which included the city, provided the rest of the money to build the rail system from a levy on sales taxes. The rail line showed signs of inequity, running through communities where it failed to stop. In one sign of race/class-based acrimony, the initials for Metropolitan Atlanta Rapid Transit Authority were translated into "Moving African-Americans Rapidly Through Atlanta." This was one more example of how different the situation can look from the grassroots perspective.

The goal is a community-building plan, not a transit plan, as enlightened transportation officials know, and existing structures get in the way. "Planning is so siloed," Scott observed. "It's getting better." The old way was for transit officials to stay within boundaries and be purely operational. The new way is to think more expansively and connect the pieces. Scott said, "You look at everything, even if it's not your direct responsibility. You need to work with others."

When entities don't work with others, and there are too many independent territories to begin with, the result is situations like those of Miami and Miami Beach, where not enough of anything connects yet. Miami Beach has bike sharing but no light rail, and its bus stops are well lit; Miami has light rail but no bike sharing, and very poor lighting around bus stops. I'll come back to the issue of connections and connection points later in this chapter.

THE BEAUTY OF THE BUS

Subways take cars and riders off the streets and operate at higher speed with mass efficiencies. But in today's densely populated cities, is it feasible to rip up a center-city road to build a subway or add rail lines? The trend is in the opposite direction, to remove tracks. Adding stations on existing lines can help underserved neighborhoods finally get more convenient access, although even that counters a trend to make things faster by eliminating stops. I say fix what we have, but choose megaprojects very carefully, focusing on connections out of the city like Washington's Silver Line to its international airport. Within cities, there are alternatives.

As public transportation goes, the bus makes so much sense. Buses are flexible. They can be replaced or moved around. They can be environmentally friendly hybrid vehicles. And buses can be cost-effective, as I've seen in my home region; a recent five-year MBTA capital plan included $1.3 billion for a short expansion of the Green Line, the granddaddy of U.S. subways, plus $835 million to replace aging subway cars and improve signaling, while a modest $75 million can purchase 342 new buses.[15]

ENTREPRENEURS ON THE BUS

Those economics make buses an attractive target for entrepreneurs. Private bus services have operated as long as there have been buses—intercity buses, tour buses, hotel airport shuttles, campus shuttle services, buses to take employees from remote parking lots to the workplace. In New York City, an informal, unpublicized, quasi-legal shadow bus system has operated since the 1980s, with direct service between the Chinatowns of Manhattan, Brooklyn, and Queens, making the trip in less than half the time of a subway. Similar "dollar van" networks, now about two dollars, serve other ethnic communities in peripheral areas.[16]

Technology enables new models for flexible, data-driven private bus services. In early June 2014, Matt George, a 2012 graduate of Middlebury College, in Vermont, began his first trial of Bridj, the prototype for an on-demand private urban bus service that could be scaled up nationally. George had done this at Middlebury, attracting venture capital funds to expand the idea to other college campuses. After college, George hatched Bridj in the Cambridge Innovation

Center near MIT, with backing from the likes of Scott Griffith, former Zipcar CEO. Bridj aims to run short direct routes to fill spurs
between main public transit lines, as Google's private buses were
doing between San Francisco and Silicon Valley. Data analytics
would determine areas of demand, and members could help determine routes and times. Griffith calls it "Uber with buses." Plus water
and password-provided Wi-Fi.

Trial service began with morning rides from Coolidge Corner
in affluent Brookline to MIT and Kendall Square in Cambridge
and downtown Boston. In early July 2014, a month after launch,
my designated rider, research team member Dan Fox, who also
rode New York's shadow vans, signed up for a free trial. He took a
public MBTA ("the T") bus from his home near Harvard Square to
Coolidge Corner, where he connected with Bridj. He reported to
me, "The T bus was typical for Boston: crowded, hot, loud, and running ten minutes late. The Bridj experience was distinctly different. When the Bridj bus (actually a shiny black Mercedes Sprinter
van) arrived, the very young driver welcomed me and the other passengers, checking our names off a list on his tablet. Seats are leather
armchairs, and there's powerful air-conditioning—pleasant on a
muggy morning. Other passengers were twenty-something professionals sporting iPhones. Doors closed right on schedule, and the
trip took about twenty minutes—a big improvement from a very
slow, jerky, and loud Green Line trip and transfer to the Red Line at
Park Street. It was really more like a car service than a bus. From
that perspective, a $5 to $8 fare seems like a good deal compared to
the $15 a cab would cost."

Just a few months later, Bridj was generating revenue that covered
per-user costs. In August 2014, high-profile Gabe Klein, whom we'll
meet later in this chapter for his work on bike sharing as Chicago
transportation commissioner, joined Bridj as chief operating officer.
At a session at Harvard in September, Matt George reported that he
has a good relationship with the MBTA, which faces increasing ridership with a constrained budget, and loves Bridj, he says, as a way

for the private sector to fill service gaps. George has even offered to share data with the public transit authority. Boston officials have been helpful, but the City of Cambridge is another story, approving Bridj's operation within the city limits but not within three-quarters of a mile of the most important places, Harvard Square (near Harvard) and Kendall Square (near MIT).

Bridj is already generating the same controversy as Google buses in San Francisco. Google uses public bus stops with shelters to pick people up and let them off. That raises numerous questions. Are public transit stops a public good, or are they exclusive to the public transit service? Cambridge responded to Bridj by banning the use of public bus stops. New transportation infrastructure often provokes discussion of what "public" means. Does public property mean city owned with controlled uses, or available for all to use in ways beyond the original purpose?

Public transit is also known as mass transit for a reason, because it is designed to serve the masses. Private services, like private schools and private aviation, are a solution for a small, privileged number, and they can remove people with clout from demanding change in the public system as they abandon it for private alternatives. George argues that Bridj could provide better and cheaper service than the MBTA in underserved areas, carrying people to well-paid suburban jobs, if it had the same subsidy. But it's also instructive that once-private urban transportation services were taken over by public authorities, which subsidize them.

Young entrepreneurs interested in urban transportation might indeed create breakthroughs. Matt George's investors might even make money, profiting from a market of people willing to pay a premium for convenience. But that doesn't make it enough of a solution for people who depend on low-cost, widely available public transportation.

This ride on an alternative model drives us right back to the heart of the issue for cities: how to reinvent public transportation systems that are stalled and emitting problems.

BUS RAPID TRANSIT: TAKING
PUBLIC BUSES TO A BETTER PLACE

Just adding buses to already clogged streets is not a solution, not to mention how unpleasant riding the bus can be in some places, especially while carrying awkward loads. This is all felt acutely in the Chicago metropolitan area, which features the seventh-highest road congestion in the country and has estimated costs from that congestion of $7.3 billion annually. Chicago's inadequate transit system ranks fifty-third in labor market access, as I've indicated, with only 22.8 percent of residents able to get to their jobs via public transit in ninety minutes or less.[17]

As Chicago officials aim to increase economic growth and access to jobs, they wonder how to reduce congestion simultaneously. "You can't grow and have congestion decline unless you start using your roads differently," Scott Kubly, Chicago DOT managing commissioner, explains. "You cannot have the same assets and use them in the same way, and have both of those things get better." Like those in most U.S. cities, streets in Chicago were built decades earlier, and they are ripe for reimagining.

Studies on one downtown Chicago avenue found that 56 percent of the people on the street were on foot. The researchers believed this was mostly because walking was faster than current buses. At the same time, 47 percent of the people traveling on roads were on buses. With nearly half of the travelers on buses, and others on foot who might prefer to use buses, Chicago leaders were ready for new models. Over a hundred Chicago Transit Authority (CTA) subway stations were scheduled to be renovated, repaired, or rebuilt, and that would keep the subway going. But for buses, planners decided to do something different, at least for America: to create a state-of-the-art Bus Rapid Transit system.

Bus Rapid Transit, familiarly called BRT, is a road system surrounding the bus that reduces delays and makes streets safer and less crowded. It combines the efficiency of rail rapid transit with the flexibility and comparatively lower cost of bus service. Its central feature is a dedicated bus lane with boarding platforms in the street, by which people can board at the same level as the door to the bus—similar to the jet bridge by which passengers board planes. They can board quickly, safely, and easily move strollers or wheelchairs. The "rapid" part is slightly misleading; it means a little bit faster, but not necessarily all that fast. We're talking going from a crawl to twenty or thirty miles per hour. Yet a few seconds of time saved here and there can add up to precious minutes and more saved per trip, keeping buses moving downtown, where they are supposed to come every two minutes, and minimizing delays felt miles away in the outer edges of the city.

This isn't new to the world, but it's new to America. BRT is well established in other nations' premier cities. Paris, Beijing, Jakarta, Mexico City, Bogota, and São Paulo have it; Curitiba in Brazil, the pioneer in BRT, made it a centerpiece of a green city in the 1970s. The United States is finally catching on; the National Bus Rapid Transit Institute at the University of South Florida is one sign of growing interest. Chicago joins Boston, New York, Los Angeles, Miami, Cleveland, and Kansas City among twenty-eight U.S. cities with an embryonic variant of BRT. For example, Boston's popular Silver Line, between South Station and Logan Airport, goes underground via the Ted Williams Tunnel, among other routes, and is technically called "mixed transit" because it operates in some places without dedicated lanes. It coexists with cars and combines some features of subways and buses, as shown in the name Line, which is more subwaylike than numbered bus routes. Confusing? Passengers just want to get somewhere; they don't care what the service is officially called.

Definitions matter because they determine which federal funding bucket a city can dip into. Not all projects in U.S. cities meet the four criteria for BRT designation. First and foremost is the separation that offers speed: a dedicated center bus-only lane running in each direction, or a separate bus right-of-way separated from other traffic. Other criteria include traffic light prioritization, which extends green lights and shortens red lights for buses; at-grade boarding from platforms, so that passengers don't have to walk up steps to board; and prepaid fares, made possible by kiosks, websites, or apps. Implementing BRT means more than painting lines on the street; some construction is required, and drivers have to change their habits.

Chicago is trying to do all of the above. The city's foray into BRT began in 2011 with an alphabet soup of government agencies—CTA (Chicago Transit Authority), CDOT (Chicago Department of Transportation), FTA (Federal Transit Administration), among others—assessing options, costs, and benefits for BRT on two main drags, Ashland and Western Avenues, while a pilot was under way on Jeffrey Boulevard. Studies showed that BRT could greatly reduce regional congestion, while also serving the underserved—those without access to rail transit.

A first test was CTA bus route J14, on the Jeffrey Boulevard corridor. A dedicated bus lane, largely separated by painted lines, was begun in the summer of 2012 and open for business by November, but it didn't yet have all the features of full BRT. Still, the Jeffrey Express bus was renamed the Jeffrey "Jump." Stand-alone Jump kiosks enabled prepayment and provided information. Whereas the Jeffrey J15 local, which still ran, made thirty-seven stops between 103rd Street and Lake Shore Drive, the Jeffrey Jump made just fifteen stops. Signaling that this was something totally new, Chicago's BRT fleet has a readily identifiable look with bright blue, Chicago's official color, adorning the bus; regular buses are a much blander white, with small red and blue stripes. Some frequent riders were skeptical of the claims. "The CTA has been promoting their new Jeffrey Jump route like it was the greatest invention short of the paperclip," wrote James Porter, a musician and DJ from the South Side, in *Streetsblog Chicago* on April 1, 2013. After riding the whole route and arriving nine minutes later than he usually did on the pre-Jump bus, he felt the Jump idea had a distance to go.

For the full rollout, Ashland Avenue was the winner. The cost would be $150 million, with multiple benefits topping that in magnitude. The sixteen-mile-long north–south road already had the highest bus ridership in the city—a whopping 10 million boardings in 2012—and the potential for significant economic impact. Trucks used Ashland to transport products from rail yards to the airport. The planned BRT line could provide access to over 133,000 jobs, reach community colleges and schools, connect to nine subway or train stations and thirty-seven bus routes, increase bus speeds by 83 percent, save commuters on average sixty-five hours each year, be available to 25 percent of the households located along the path that do not own a car, and increase safety. Businesses located near the BRT line would have increased visibility and potentially more visitors. Faster, more efficient travel would create easier access to jobs. Increased ridership would make it profitable for the transit authority. Economic development benefits made it possible to use tax-increment financing, in which additional property tax revenues from increased values along the route could pay for transit projects— the way railroads had been financed in their early days.

You can't please all the people all the time. As the BRT concept was discussed at civic meetings and community forums, a litany of concerns emerged that will sound familiar in any community and around any new project. Some are against anything that involves taxpayer dollars. Others don't like the specifics: the elimination of a vehicle lane in each direction, and the prohibition of some left turns. That's classic resistance to change: the work involved in learning new routines. Another group complained about the proposed increase in distance between stops (every half mile compared with current bus stops at every one-fourth mile). Small businesses worried about BRT's impact on their day-to-day operations. Some considered it a bad idea altogether because of Chicago's severe winters—proof that people will stick with a painful status quo rather than risk change.

"Because we had to remove the right turn lane and take a lane to convert it to BRT, we spent a lot of time explaining how traffic would flow better for everyone," Keith Privett, a Chicago DOT planner noted. "There was a lot of collaboration and coordination with the transit authority to come up with new designs within the existing layout of streets. The toughest challenge was satisfying businesses and their need for loading and unloading."

All the listening and educating worked. Supporters of BRT soon well outnumbered detractors. Citizen polls in May–June 2013 underwritten by the Rockefeller Foundation showed 59 percent of respondents in support of BRT and only 10 percent opposed. Nearly three-quarters of respondents said they would switch from driving to a faster alternative like BRT—a huge number even if exaggerated, because people always have better intentions on surveys than in reality. Equally impressive was that BRT plans had over thirty official and diverse supporters, ranging from the Illinois Medical District to Blue Cross and Blue Shield of Illinois to the Chicago Children's Museum to the Illinois Hispanic Chamber of Commerce. BRT champions included a who's who of civic organizations: the Metropolitan Planning Council, Chicago Architecture Foundation, Active Transportation Alliance, Urban Land Institute, Civic Consulting Alliance, Institute for Transportation and Development Policy, and Chicago Community Trust.[18]

In May 2014, I stood in the path of the Central Loop BRT, a project financed by a $24.6 million Federal Transit Administration grant and $7.3 million in local Tax Increment Financing bonds.[19] A rotating group of officials and staff from the Chicago Department of Transportation led a walking tour of the main downtown boulevards that went through the heart of the Loop's high-rise office buildings to the Chicago River and across to Union Station.

Planners Mike Amsden and Keith Privett, along with Chicago DOT assistant commissioner Sean Wiedel, sang the praises of dedicated lanes in the middle of the avenues. The cost and disruptions for construction were much less than would be the case with other new infrastructure options. We saw dug-up portions of streets with deep holes, even though the bus lanes were all aboveground. The city was repairing and upgrading sewers and water mains first, so that difficult infrastructure work would be out of the way before work on the bus lanes began.

As we walked the crowded sidewalks before lunchtime, we could easily see what was wrong with the current system, and why BRT would be an improvement. Buses were slow, and they slowed down everything else. Cars competed with buses for lanes, especially cars turning right, and would block curbs. Trucks unloading from the curb would also keep buses from moving. Buses had a small area to pull into at the curb to pick up passengers. Passengers took up space on the sidewalk while waiting at a bus shelter, forcing pedestrians to squeeze by. We saw the problems compounded, as cars were stuck behind a bus that was blocked by a delivery truck.

What about Federal Express lanes? I asked my hosts, only half facetiously, spotting the FedEx truck parked across the street. Or water bottle delivery trucks? Should they have dedicated lanes too? There was polite laughter and a serious answer. Planners were working on trying to get companies to make deliveries early in the morning, and they wanted vans to use alleys if such pathways existed. "The ultimate goal is to try and make the streets work better for everyone," Mike Amsden said.

CONNECTION POINTS

What ultimately makes buses and subways desirable to the public is not simply that each route is faster but that they connect faster and more conveniently with one another and with all the various ways people go places. That requires physical connection points. That's why cities are eyeing their aging central train terminals, often a century old. That's why my walking tour group was eyeing Chicago's Union Station. Aboveground, it is a mammoth set of buildings spanning several city blocks surrounded by traffic jams. Belowground, it is a cavernous structure, crowded with food court users in one area (our lunch stop) yet completely empty in the grand marbled concourse. Not surprisingly, it is on the high-priority-for-change list.

Just as Mayor Rahm Emanuel was taking office, Chicago DOT, Amtrak (the station's owner), Metrarail (the station's primary tenant), with others, began to envision reinventing the main train terminal in the heart of downtown as a new intermodal connector, where BRT meets commuter rail and long-distance trains. Chicago's Union Station is the third-busiest railroad terminal in the United States, serving over three hundred trains per weekday, carrying about 120,000 arriving and departing passengers—a level of passenger traffic that would rank it among the ten busiest airports in the United States. About 90 percent of the people flowing through it are taking Metra commuter trains, some of them connecting with buses. The station is also the hub of Amtrak's network of regional trains serving the Midwest as well as most of the nation's overnight trains, which connect to the Atlantic, Gulf, and Pacific coasts. It's not just that the station often operates close to capacity because of a revival of interest in rail; it's also a mess that ties up street-level traffic—taxis, passenger car pickup and drop-off, shuttle buses, CTA buses, an unusable bike lane marked by barely discernible fading paint. Entry points are difficult to use. This former Chicago landmark opened in its present form in 1925. In 1970, significant alterations were made to accommodate more trains. That's basically it. Nearly forty-five years of neglect, and only in the 2010s is it getting renewed attention.

Washington's Union Station started earlier (1907) and is further along. By the late 1970s, mold-filled sections of it were deemed to be in imminent collapse, even as a subway system was being planned for the nation's capital as an alternative to the Beltway highway project. In 1988, Secretary of Transportation Elizabeth Dole worked on funding for renovations and better connections. Now it is a retail and food destination for pedestrians as well as a major connection point for Amtrak long-distance trains, including the Acela Express; MARC and VRE, the Maryland and Virginia commuter rail lines; and the Metro subway system, which connects to Reagan National Airport and, at Union Station, has direct access to the street. In 2011, a bus facility was added for intercity bus lines as well as the DC Circulator city bus, which offers a bus every ten minutes during operating hours. Just about every mode of transportation and a connection to nearly everywhere in the city or the world is under one roof. There is still much to do. Amtrak is looking for $7 billion in funding for a fifteen- to twenty-year expansion plan. A Metro extension to Dulles Airport is under way in two phases. And although people can connect by walking, the systems are not digitally connected. It's a start, however, and has contributed to quality of life and car-free transportation.

Denver's Union Station opened its new facilities in the spring of 2014, around the time when I was walking through downtown Chicago. A group of my Harvard fellows toured it in June. It's designed not so much for current connections as for a public transportation future. Denver is a frontier town that grew up on the automobile; only 6 percent of Denver area commuters use bus or light rail—half of the transit mode share in Los Angeles, for example. FasTracks, the public transit system that will connect there, has been called "the most ambitious plan for expanding public transit in the United States." Transportation investment is spurring neighborhood redevelopment. As of mid-2014, the LoDo (Lower Downtown) redevelopment effort has created 7 million square feet of new office space, 5.5 million square feet of retail, and 27,000 new apartment and condominium units. The project also has innovative governance: a multijurisdictional Denver Union Station Project Authority—a nonprofit, public corporation that could bear the project's debt load and work with private investors as a public-private partnership. Judy Montero, a Denver city councilwoman, calls the organization "a phenomenal example of regional partnership," bringing together often squabbling jurisdictions with a "common vision" for the region. Colorado DOT executive director Don Hunt promises, "It's going to be flexible enough to transform itself over the decades as technology changes with transportation."[20]

Transit stations must transform themselves for other reasons. People arrive for high-speed services by much slower, much older means, on foot or on bicycles.

LEGWORK: POWERED BY THE PEOPLE

The ultimate low-cost, green, convenient, personal vehicle for mobility is ourselves, aka walking. Close behind and capable of longer distances at higher speed are bicycles, which are also people powered. Bicycles are an urban solution for work, not just fun, that resonates with the values of new generations flocking to cities.

In both Germany and Denmark, more than 20 percent of commuters travel by bicycle. In some Chinese cities, cycling still accounts for more than half of all trips, although rates are dropping. But the Netherlands qualifies as the bicycle capital of the world. Nearly 30 percent of Dutch commuters always travel by bicycle, and an additional 40 percent sometimes bike to work. The interesting thing is that biking gets less risky when more people bike. As more cyclists fill Dutch streets, bike fatalities have remained among the lowest worldwide and may actually have declined. From 2002 to 2005, an average of 1.1 Dutch biker was killed per 100 million kilometers cycled. There's safety in numbers.

In the United States, fewer bicycle users mean more danger. From 2002 to 2005, with less than 2 percent of trips in North America by bicycle, fatality rates are 5.8 American deaths compared with 1.1 Dutch per 100 million kilometers cycled.[21] It's not just whether there are bike paths or lanes; U.S. roads are dangerous altogether. In 2012, traffic injuries on the nation's roadways claimed the lives of 33,561 people, higher than the 2011 death toll of 32,479. That's a 3.3 percent increase—a difference of more than a thousand lives—and the first time since 2005 that the number of fatalities has gone up.[22] What's critical for cities is that pedestrian and bicyclist deaths rose faster than the overall rate, by about 6.5 percent each. In 2012, 4,743 people were killed while walking and 726 while biking. Overall traffic injuries rose 6.5 percent—but 10 percent for people walking. Cyclist injuries went up by 2.1 percent.

Wheels are revolving. Biking to work has increased 60 percent over the last decade, the U.S. Census Bureau reports. The National

Household Travel Survey showed that the number of trips made by bicycle in the U.S. jumped from 1.7 billion in 2001 to 4 billion in 2009. Between 2011 and 2012, bike commuting grew 9 percent to include almost 865,000 people, although that's just a minuscule 0.61 percent of the commuting U.S. public.[23] In some cities, the numbers are creeping up. In 2012, the ten major cities with the highest percentage of bike commuters were Portland, Oregon—at 6.1 percent, well ahead of the pack—followed by Minneapolis, Washington, D.C., Seattle, San Francisco, Denver, Tucson, Oakland, Sacramento, and New Orleans. The list shows how universal bicycle commuting could become. Bad winters don't seem a deterrent, or Minneapolis and Washington, D.C., wouldn't make the top list; nor do steep hills discourage cyclists in San Francisco and Denver. Intrepid cyclists pedal on regardless.

However much the growth, these are still tiny numbers. In mid-2014, Knight Foundation vice president for community and national initiatives, Carol Coletta, created a competition to get U.S. cities to be more aggressive about bike friendliness and took delegations from the winning cities—Charlotte, Detroit, Lexington, Kentucky, St. Paul, and San Jose—to Copenhagen for intensive study and inspiration. Detroit was chosen, she told me, because it is starting over and has a chance to get it right; San Jose, because it has an ambitious goal to increase dramatically the percentage of people commuting to work in other than single-occupancy cars.

Change requires more than infrastructure, such as bike paths, lanes, and traffic signals. A culture of multimodal streets must develop, and vehicle drivers must respect bicycles and pedestrians. Safe infrastructure for bicycling is becoming a priority in cities across the country. Cities like New York, Chicago, and Washington, D.C., are competing to add the most bike lanes. Smaller cities like Indianapolis and Memphis are also building bike routes at an impressive rate. In 2012, the number of protected bike lanes in the nation rose from 62 to 102.[24]

The rapid growth of bike sharing is reducing barriers to cycling. Here, too, America is behind. Amsterdam started the world's first bike-sharing program in the summer of 1965.[25] Called the White Bicycle Plan, it almost toppled before it took off. Its free bicycles were supposed to be used for one trip and then left for someone else. Within a month, most of the bikes had been stolen, and the rest were found in nearby canals. There are always start-up pains. But then the concept gained traction. Now it's a well-established urban amenity, maybe even a social movement.

Bike-sharing programs are spreading across the United States, with shared bikes in at least thirty-six urban areas, including Boston, Chicago, Denver, Fort Worth, Miami, and New York City by 2014, up from just six programs in 2010. Alta Bicycle Sharing, led by a former bike planner for Portland, Oregon, runs programs in Chicago, Washington, San Francisco, New York, Boston, Columbus, Toronto, Chattanooga, and Melbourne, Australia.

New York City launched its bike-sharing program on May 27, 2013, for Memorial Day weekend, without using any taxpayer dollars. This was one of many green initiatives envisioned by Mayor Michael Bloomberg, who had a transportation focus. It was entirely funded by Citibank, which paid $41 million for six years, with no operational responsibility for what became Citi Bikes, a brand so strong that it discouraged other potential sponsors. Since its rollout, under a contract with Alta, the Citi Bikes system has been plagued by maintenance and software problems. Alta, a small company with limited resources, has had difficulty raising capital to improve its operational issues and expand throughout the city. Revenue targets weren't met because lower-cost commuter ridership exceeded expectations, while high-cost/high-profit day-use memberships did not sell well.

The city prevented Alta from raising annual membership fees without first finding a financial backer. Finally, REQX, an affiliate of the real estate firm Related Companies, acquired a majority stake in Alta and provided cash to double the Citi Bike fleet size to about 12,000, expand bike sharing into Queens and Brooklyn, and fix operational issues. It was stop and go there for a while, with deal approval stalled. But on October 28, 2014, a deal was announced that will expand the program to support over 700 stations and 12,000 bikes, extend the program's reach in Brooklyn and Harlem, and bring bikes to Queens.

In Chicago, bicycles had been on the city's agenda since the 1990s. Between 1995 and 2008, federal Congestion Mitigation and Air Quality grants supported a series of Chicago Streets for Cycling bikeway projects that included approximately ninety-four miles of on-street bike lanes and twenty miles of marked shared lanes. In 1992, a Bike 2000 plan was created. Beginning in 2001, the city deployed a small group of bicycling ambassadors and safe routes ambassadors for pedestrians as educational resources for schoolchildren. In 2005, the city council adopted a Bike 2015 plan with two big goals: reduce the number of injuries by 50 percent from 2006 levels and increase bicycle use, so that 5 percent of all trips of less than five miles would be by bicycle.[26]

Everything accelerated when newly elected Mayor Rahm Emanuel brought Gabe Klein from Washington, D.C., in early 2011 as commissioner of transportation, the top Chicago DOT official. Klein is a one-person embodiment of the new values-laden, sharing economy-promoting, local-food-eating urban millennial. After graduating from college in 1994, he joined Bikes USA, one of the nation's largest bicycle retailers, rising to director of stores. From 2002 to 2006, he was a regional vice president of Zipcar, overseeing car sharing in Washington. He later cofounded On the Fly, an electric vehicle-based, natural-local food stand company whose vehicles could work on the road or sidewalk. In 2008, a new D.C. mayor lured him from the private to the public sector as director of transportation; there he developed and implemented Capital Bikeshare, a pioneering U.S. bike-sharing program, during his two years in office. And in 2014, after leaving his Chicago post, he became part of the urban private bus story by joining Bridj.

Upon his arrival in Chicago, Klein immediately set new priorities. Henceforth, transportation projects would put pedestrians first, followed by transit users and cyclists. Cars were at the bottom of the hierarchy. CDOT wanted to eliminate pedestrian and bicyclist deaths and reduce total crashes and injuries in the city by 10 percent every year. In addition, parking spaces would be converted into "people spots," where people could enjoy a minipark, some art, or a performance, as part of the placemaking movement, which engages the community to reimagine the use of public spaces. The goal was "enriching the experience of the community and bringing back a sense of pride and vitality," MarySue Barrett explains. She is president of the Metropolitan Planning Council, which provides technical assistance to community groups for placemaking (and does much more for the region), while the city invests in cross-department training about it.

These concepts are wrapped together under the umbrella of Complete Streets, a coordinated, integrated approach aimed at fundamentally reinventing streets. Bicycling and bike sharing are a major emphasis, along with BRT and other public visions. Divvy was chosen as the name for Chicago's bike-sharing program, because it was unique, fun, and suggested sharing. Divvy it up.

You'd think getting bike sharing under way would be a no-brainer. Bicycles are healthy, low-cost, green, democratic, and benign. Bike sharing is an enlightened civic thing to do. So just do it. But think again. From the beginning, headaches developed in Chicago. Some were logistical: finding enough space on streets and sidewalks for cars, trucks, buses, pedestrians, and bikes, identifying places with solar power to run the technology, and then developing good places for bike stations. High density is ideal to ensure high utilization, which is necessary for financial sustainability, but by definition there's less space.

The logistics for bike sharing are easy compared with the politics. People whined that a $65 million contract was awarded to Alta, the same company operating New York's Citi Bikes, rather than a cheaper local bidder, or that this was not a core government service, especially because the city was in the midst of a fiscal crisis. A condo association in an affluent North Side neighborhood went to court to remove a Divvy bike station, saying stations didn't look good, crowds would be unsafe, and property values would decrease.[27] But others complained about the opposite—why weren't they getting a Divvy station, with its convenience and value?

Fights broke out. A city alderwoman proposed a $25 annual cycling tax, like a registration fee for cars, to help pay for the Divvy program and bike lane snow-clearing services. Some motorists objected to the coddling of cyclists on roads paid for by a gas tax that cyclists avoid ("stop-sign-running freeloaders," a news account called bicycle riders). Cycling advocates warned of possible oppression by bike cops who could not enforce a tax anyway. A blogger suggested that pedestrians pay a shoe tax to use the sidewalk. Can't we all just get along?

After patiently seeking input from diverse stakeholders (another familiar action) and collaborative work across city departments and subdepartments, Klein's team launched Divvy. Klein waxed eloquent about Divvy as a cost-effective transportation solution, telling us, "From a business standpoint, from the government's standpoint, the ability to literally drop a station in twenty to thirty minutes, no construction, no wiring it to the power grid, and then being able to expand it on the fly was like nothing else out there in transportation." There are other nifty features. Technology enables easy tracking and control of system use and transactions. Bikes could potentially contain environmental sensors, gathering information for city planning and disaster preparedness.

Start-up and implementation costs were around $22 million. About 80 percent was covered by a federal congestion mitigation grant; bicycles are a pollution solution. Additional funds were raised from the City's Tax Increment Financing program, the same program that was helping fund Bus Rapid Transit for Ashland Avenue, on the grounds that later tax revenues from enhanced property values in areas with bike-sharing stations could be used to repay the debt. Divvy was designed as a profit-maker once it got started, so usage fees had to cover operating costs and more.

Divvy officially launched on June 28, 2013, with 69 stations, 1,500 members, and about 2,000–3,000 trips per day. By October, a full system of 300 stations and 3,000 bicycles was in place. Growth was rapid. As of May 28, 2014, there were 19,000 members and 7,000–9,000 trips per day. On May 21, 2014, Divvy hit 11,000 trips. On Sunday, May 25, during a beautiful Memorial Day weekend, Divvy reached a record of over 21,000 trips. Two months later, in July 2014, Divvy had swelled to over 21,000 annual members and even more casual users, 212,000 of them buying twenty-four-hour passes, 65 percent of those to regional locals and the rest to visitors.

In late May 2014, on my Chicago rounds, I stood with a group by the Divvy bike station in front of 30 N. LaSalle, a high-rise office building, catty-corner from Chicago City Hall, including the officials and planners with whom I was discussing Bus Rapid Transit, Sean Wiedel, Mike Amsden, and Keith Privett. They were showing off Divvy when avid cyclist and director of planning services Jeffrey Sriver joined us.

Divvy stations announce their presence with a bright blue color, the official color of the city, used also on BRT buses and, conveniently, the color of Divvy's first private company sponsor, Blue Cross/Blue Shield. We could see it on the Chicago flag nearby. At 11 a.m., all the bike-docking slots were full, but at commute times and lunch, the stations empty out fast. If one is empty, another is just a block or two away, in a dense downtown array dotting Central Loop streets. I could also see many non-Divvy cyclists on the streets even in the quiet prelunch hour.

The CDOT planners ride their personal bicycles to work but use Divvy too. Wiedel and Amsden say that they find it easier to leave their own bikes in the storeroom in their building and then use their Divvy keys to grab a Divvy bike for getting around town. The system is geared to thirty-minute bursts. They are also part of the new urban wave experiencing the new urban lifestyle. Sriver, who has a graduate degree from MIT, has worked all over the world. He jokes about how his work on transportation projects throughout the Third

World prepared him to come back to America, which needs some of the same fixes. Toward the end of my visit, we drove past his neighborhood. He lives in a town house complex on the edge of downtown in what had been a rail yard, not far from what had been a train station. He walks his middle school daughter a few blocks to school and then bikes to work. The family goes biking on weekends down city streets, past the old rail landmarks. Dearborn Station is now a shopping mall—Chicago future rising out of Chicago past.

Changes to create a bike culture are already in place on the wide avenues and boulevards that were about to be modified to carry BRT buses. Bike-only lanes are placed so that they do not bump against traffic, and parking for cars is toward the middle of the street rather than beside curbs. Signs and traffic signals educate cyclists and motorists alike in bike etiquette. Bike lanes are marked with a bicycle symbol, with green or red signs making it clear that bikes should stop. This is common in bike-friendly cities in Europe but new to Chicago and most U.S. metropolitan areas. This means that taken-for-granted routines must be relearned. Sociologist Erving Goffman once pointed out that it's miraculous that we can confidently step off a curb into the street without fearing injury because we know that cars will stop at traffic lights (well, at least most of the time), because the development of a car culture shaped expectations. Cities now need a similar set of norms for bikes.

There are signs that Chicago's emergent bike culture is making a difference on at least one important safety indicator. Chicago recorded a tragic forty-eight deaths of pedestrians and cyclists on city streets in 2012. The number was down to twenty-six in 2013, the year after bike infrastructure was put in place and the year Divvy was launched.[28]

Bike sharing might change the culture, but it doesn't reach everyone yet. We're back to transportation disparities. Chicago's Divvy stations are placed on the basis of anticipated volume so that the program can pay for itself. That translates to every few blocks downtown, about every quarter mile in the wealthier, predominantly

white North Side, about every half mile in the predominately black
South Side, and even farther between in poor, predominately Latino,
neighborhoods on the West Side. The annual fee requires a credit
card, creating a further barrier for people without bank accounts.

Bicycles can go a long way to remove cars and move people, but
they won't create opportunity unless tied to more comprehensive
visions for change.

AUGMENTING FOOT POWER WITH TECHNOLOGY

Bicycles are claiming their place on city streets. But what if not
merely streets but personal vehicles themselves could be reinvented?

Segway offers an instructive lesson in why some dreams don't come
true. Just because innovators and entrepreneurs can envision engineering marvels doesn't mean people will want them (or want to pay
the price). Dean Kamen is a prolific inventor who grew wealthy from
health care inventions such as the first automatic IV dosage machine
and a portable dialysis machine. He later found his way into transportation, which he saw as a sector that had been stuck for decades, if not
centuries, without innovation in personal mobility. While working on
a self-balancing, stair-climbing wheelchair, he came up with the Segway as an alternative to bicycles, yet small enough to stay on the sidewalk rather than riding on dangerous roadways.

Segway was code-named Project Ginger and kept under wraps,
which built buzz about the mystery of what in the world Kamen was
doing. I heard speculation about Ginger when Kamen spoke at the
World Economic Forum in Davos, Switzerland, where I met him.
When Segway was revealed in 2001, Kamen said that his invention
fulfilled an unmet need for speedy, green urban transportation and
would be "to the car what the car was to the horse and buggy." Legendary venture capitalist John Doerr predicted that Segway would
be the fastest company in history to reach $1 billion in sales as it built
a market of "empowered pedestrians."

Despite the hype, it is still just a scooter. After its 2002 launch,
Segway tooled up to produce 40,000 units per year at about $3,000

each, but only 23,500 had been sold by September 2006, when the
company recalled all its models, owing to a software glitch. In 2009,
following lackluster sales and significant financial losses, Kamen's
Segway company was sold, then sold again in 2013.[29] Although there
are continuing commercial sales—e.g., for police, airport security
guards, or warehouse workers—Segway has become a novelty, available for daily rentals by tourists in authorized Experience Centers
in Chicago and Miami Beach. But Washington, D.C., banned Segway
tours in 2010, citing safety concerns for pedestrians. Even though
Norway legalized the riding of Segways on public paths in May 2014,
many U.S. cities still restrict their use on sidewalks. Without sidewalk access, Segways are curbed.

It seems that the Segway was not destined to be a game changer
after all. Skittish public officials, the same bunch restricting Google
buses and parking apps, don't want to get ahead of the people. Walking and biking work well enough to provide individual mobility
within the classic categories of streets, sidewalks, and bike paths.

Of course, there is still fertile ground for innovation in these
modes. Plenty of entrepreneurs are ready to envision smarter walking, such as FitBit and similar connected pedometers linked to
smartphones or smartwatches, which can monitor every bodily
function. Others are targeting smarter cycling, such as technology-
enhanced bicycles. One embryonic example is the Copenhagen
Wheel, created at the MIT SENSEable City Lab and commercialized
by the start-up Superpedestrian. The City of Copenhagen had chal-
lenged Assaf Biderman's MIT team to work on bike innovations, and
this is the result. The Copenhagen Wheel substitutes for the rear
wheel of a conventional bicycle. It holds a small 250-watt electric
motor; removable batteries; processors; sensors for motion, force,
and air quality; a wireless radio; and a wheel lock.

Unlike electric bicycles, the wheel supplements human power
only when needed, multiplying the bike's own momentum and also
storing energy when not in use, thus operating in a more sustainable
way. It can also serve as a moving sensor collecting data to transmit;

for example, reporting on air quality as the wheel rolls through a city, which is certainly a bonus for communities. Superpedestrian claims that adding the Copenhagen Wheel facilitates fast, easy biking without the sweat—literally. Perhaps users can arrive at work after cycling there without having to change their clothes, thus removing a disincentive for commuting by bike. It's too soon to say whether this will catch on. With thousands of preorders, the wheels are scheduled to be delivered by the end of 2014. FlyKly, a New York competitor, is targeting the same opportunity and $800 price tag.[30]

The high cost of enhanced bicycles is one more factor dividing income groups in America. Copenhagen Wheels and their cousins might stimulate professionals in suits to drop cars in favor of bicycles for short distances, but public transit is still needed for large numbers of people. Those who can't afford plain old bicycles can't even imagine innovative alternatives. And they won't be found at Divvy bike stations. Still, to some bicycle-commute and bike-sharing proponents, addressing social inequities is as much a part of their movement to rethink cities as is environmental sustainability. There's a social activism component to bike culture in some places. Bikes Not Bombs, a nonprofit founded in 1984 in a diverse urban neighborhood in Boston, uses bicycles as a tool for social change by teaching inner-city kids to recondition and repair them, with vocational education credits, and to ride them safely, thus increasing their mobility options, both economic and geographic.

Cities are on the cusp of change. But change is not a matter of a series of one-off innovations, mode by mode; it requires connection to a coherent, integrated strategy. That's what Chicago's Complete Streets idea is trying to convey.

LEADERSHIP: CONNECTED STRATEGIES

Change moves slowly, at bicycle, not rapid transit, speed. Big projects can take longer to get approved than to implement, and opposition is a given. But cities such as Chicago demonstrate that transportation and infrastructure problems can be tackled by strong public-sector leadership with a strategic vision, with the support or active partnership of the private sector. At a time when national can be ugly, local is beautiful. Cities are moving despite gridlock in Congress, and mayors are getting things done.[31] Not alone, of course. The best central-city mayors must forge alliances with governors and other jurisdictions in their metropolitan region and work through cross-sector multi-stakeholder coalitions to clear the way for big investments.

Cities at the top or rising have had mayors who became respected national figures because of their combination of pragmatism and innovation. Michael Bloomberg in New York (2002–13), Thomas Menino in Boston (1993–2013), Manny Diaz in the City of Miami (2001–09), and now Rahm Emanuel in Chicago, elected in 2011.

Chicago has been getting results and acclaim for moving while the nation seems stuck. Long known as the second city, even after Los Angeles passed it in population in 1984, Chicago is a crossroads for North America. It has the world's second-largest airport by traffic (O'Hare), the second-highest concentration in the United States of Fortune 500 company headquarters, and is the second-largest for postsecondary education slots. It is also second to New York as America's top global city in future competitiveness by 2025, according to a ranking by the *Economist* in June 2013, having moved up three places, from number 12 to number 9, since the preceding year. Chicago earned first place globally for its governance of environmental issues, also receiving high ranking for improvements in government effectiveness, quality of health care, flight connectivity, and public transport—the result of projects we have just toured. And here's a dubious number 1, but a status made possible by the attractiveness of the city to conventions and tourists: in 2013, the Global Business Travel Association ranked Chicago number 1 in highest total travel tax burden per visitor per day.

Chicago has a history of long-term, strong mayors. Richard J. Daley served as Chicago Mayor for twenty-one years, from 1955 to 1976, and later his son Richard M. Daley for twenty-two years, between 1989 and 2011, when Rahm Emanuel took office. A former congressman, Emanuel had also been President Obama's chief of staff. Mayor Emanuel inherited fifty aldermen on the city council, thirty-three administrative departments, countless civic activists of all persuasions, an engaged business community that met regularly for breakfast, and, despite progress in chipping away at long-term projects such as rail decongestion in the preceding decade, a long list of deteriorating institutions and aging infrastructure. Chicago received a D minus on the American Society of Civil Engineers report card for infrastructure. Emanuel also inherited major financial and human challenges, such as an unfunded pension liability of $19.5 billion, a high homicide rate, primarily in impoverished areas, an ailing K–12 public school system, a lagging community college system, and infrastructure projects still waiting to be finished.

In Chicago, Emanuel quickly unveiled a vision to bring Chicago into the twenty-first century from a twentieth-century mindset. In 2009, Chicago had received $1 billion in federal stimulus funds for projects, including rebuilding forty-three miles of pothole-filled arterial streets, the kind of shovel-ready repair work that's necessary but doesn't build the future. Emanuel's Building a New Chicago plan offered a comprehensive strategy involving a high level of coordination between the city's departments, private sector, and the public, to improve the streets, water system, schools, parks, and transportation. It was a can-do and can-do-it-now plan. The estimated $7 billion in investments did not have to wait for federal assistance or require raising taxes; funding would come from efficiencies realized, office cuts, water fees, new revenue sources (such as licensing to put digital billboards on city land along highways), and an envisioned first-in-nation Chicago Infrastructure Trust (more about that in the next chapter). Mayor Emanuel's transition committee included a transportation subgroup staffed by the Metropolitan Planning Council, MarySue Barrett's organization. World Business Chicago, a collaborative of business and civic leaders, chimed in on an economic plan and also handled complex projects that lacked a clear leader and/or financing. Public-private collaboration all the way. And, of course, federal funds were tapped, and bonds issued, including the tax increment financing that helped with Bus Rapid Transit. In 2013, Emanuel announced that the city would be exchanging land along highways for millions of dollars in revenue from digital billboard advertising, a creative move but criticized by safety advocates, as I've pointed out.

To avoid getting stymied by silos, Emanuel did more than just insist that departments work together; after all, talk is cheap. He did some reorganizing and created new chiefs within his office who could integrate efforts across departments, such as a chief technology officer, former IBM director of citizenship and technology John Tolva, and a chief sustainability officer, former senior vice president of Shore Bank and a coproducer of the climate change film *Carbon Nation*, Karen Weigert (also a Harvard Business School M.B.A. and my student).

An integrated technology plan unveiled in September 2013 featured twenty-eight initiatives organized under five broad strategies, including next-generation infrastructure and making every neighborhood smarter. By the end of 2013, eight parks and beaches were on track to have Wi-Fi. The city had nearly completed gathering the inventory of unused municipal assets—such as sewer lines—that could be repurposed for the commercial gigabit networks. Chicagoland Chamber of Commerce president Theresa Mintle enthused, "Business loves the technology plan, because broadband is as important a part of the city's infrastructure as the electrical grid was a century ago." To ensure skills for a technology-enabled world and improve educational prospects in a troubled system, a definite two-fer, Emanuel started College to Careers soon after he took office in 2011, a program that pairs local schools with employers to better prepare students for the workforce. He then leaped into innovation by personally recruiting first IBM and then Microsoft, Cisco, Verizon, and Motorola to each partner with a high school and a community college to create five STEM (science, technology, engineering, and math) early college high schools, which opened in September 2012 and showed immediate results. These schools used the model IBM had designed and launched in New York City with enthusiastic support from Mayor Bloomberg the preceding year, as Pathways in Technology Early College High School (P-TECH).[32] The model is a striking innovation: an open-enrollment six-year high school that offers any student who attends the chance to take college-level courses early, get mentoring from a tech company, and a promise of a job interview after graduating with both a high school diploma and an associate's degree.

With a fiscal crisis and scarce resources, creativity is essential. The city demonstrated kaleidoscope thinking—my term for shaking up old assumptions to find new patterns. Old things were turned to new uses. Underground freight tunnels historically used for coal deliveries were retrofitted with fiber-optics to improve broadband throughout the city. The Bloomingdale Trail project (2.7 miles long) transformed an unused elevated rail line into a green space for pedestrians and cyclists, similar to New York City's High Line, another Bloomberg innovation borrowed by Emanuel. A virtue of collaboration across departments was the ability to get multiplier effects: several things done with one stab of the shovel. When streets were opened for maintenance, another department could re-create the street to accommodate multiple modes of transportation or add fiber underground to increase broadband connectivity—or both. Complete Streets were also flexible streets. The Fulton Market area was historically a meatpacking district, but as restaurants, a new transit center, and Google moved into the area, the streets had to become flexible and adapt. The city was exploring making it partially pedestrian in the evenings, while maintaining its industrial use between 4 a.m. and 2 p.m.

CITIES ON THE MOVE:
WELCOMING PEOPLE TO THE STREETS

The daunting problems that plague most U.S. cities can't be solved quickly. As the saying goes, thousand-mile journeys begin with single steps. But unless the steps head in the same direction, guided by coherent road maps, they will drain energy without producing results. Moving a city is more complex than transforming an organization, but it takes the same basics of leadership for change: A compelling vision that connects actions to a clear set of goals. New organizational approaches. Involvement of many stakeholders to get buy-in. Leveraging scarce resources by choosing projects that accomplish many things at once. Finding quick wins that build support for much-longer-term work. Taking advantage of innovations, including new technology. Building new partnerships that bring additional expertise.

The challenge of rethinking cities is to understand the need for new connections, so that multimodal people can get where they need to go safely, quickly, conveniently, sustainably, and with opportunity at their destinations. Then when people take to the streets, streets will be welcoming. This is a new vision for cities that can usher in a new transportation era for the nation, and, in fact, solve other problems for America's future.

Cities can become greener in terms of both carbon emissions and cash. As the image of central cities changes from blight to delight, population growth is creating a new set of expectations. Could a new set of cultural icons follow, replacing dreams of a suburban white picket fence with dreams of public park views from a high-rise apartment? That's good for central, affluent, and primarily white areas, but it doesn't necessarily reach everyone. Low-income and minority communities often do not experience the benefits of improving cities, and are sometimes pushed out of gentrifying urban cores into old inner suburbs, where they lack connections to a revitalized economy. Inequities must be addressed as we build the future.

Public transit is a vital, efficient circulatory system for cities, connecting people to jobs, education, and all that a city has to offer. Without adequate transit for the less-advantaged, opportunity will tend to pass them by, like overcrowded buses on a cold winter day. Subways are valuable, but constructing many more of them would be expensive. Flexible alternatives are growing and hold promise. Bus rapid transit systems could help, as they are starting to in Chicago and other cities. Existing assets can be put to new use. Old transit hubs, like the Union Stations of Chicago, Denver, and D.C., can serve new purposes as modern connectors, and revitalize their neighborhoods in the process. And adding technology connections to existing transit makes it more accessible and transparent to its users, allowing them to optimize their trips. With a better system, maybe Martin Johnson could cut in half his laborious commute to school.

Bicycles and bicycle sharing could reshape the urban experience, but just 2 percent of U.S. commuters bike to work. With only a small number of protected bike lanes in major streets nationwide, rules and cultural norms of respect for bicycles and pedestrians are still undeveloped in most cities. Without subsidy and special efforts, bike sharing will not make a difference in underserved neighborhoods, and it will always be limited in range. Technological alternatives to people-powered pedaling are also limited in their potential for widespread impact. Privately operated buses can fill some gaps, but they can't solve enough mobility problems. Historically, private transportation systems have eventually needed public support and have been taken over by government transit authorities, with the recognition that longer-term social benefits outweigh immediate operating losses.

In the end, we keep coming back to public mass transit, whether subways, light rail, buses, or bus rapid transit, to connect parts of a city to one another, cities to the airports and the world beyond, and people to opportunity. An upgraded system of public transit is essential for mobility, and even, as we have seen, for upward social mobility. Without it, the streets fill up, and everything slows down. With it as part of the mix, cities can achieve the quintuple wins: safety, efficiency, convenience, sustainability, and economic growth.

Reimagining cities and city streets involves finding a balance between cars and other forms of mobility; cars, trucks, buses, and people must share the space in better ways. Of course, anything that disturbs local patterns will be met with local controversy, even the benign bike. And mass transit is not cheap; it will take strong leadership to overcome reluctance to invest. Fortunately, mayors have led the charge, recognizing the centrality of mobility to their cities' futures. But even this strong leadership is not enough. Mayors rely on a wide range of civic associations and civic leaders, including business leaders and nonprofit heads, who contribute to plans, rally their constituencies, and often prod the mayors, especially as activity covers a metropolitan territory wider than the central city.

Localities can't gather the resources to make changes alone, and they shouldn't. Cities are strategically important to the nation, and federal programs with federal funds are a critical part of the mix, but they have to become more strategic. To power the revitalization of America's cities, it will take vocal support from cross-sector, multistakeholder coalitions to build awareness and consensus—to convince others that everyone needs to care and that we need to find the money.

6

THE WILL AND THE WALLET

The Bridge to Nowhere is often invoked as an example of all that is wrong with the U.S. pork barrel system of allocating public funds to infrastructure projects. It's easy to see why. The Gravina Island Bridge was a proposal before Congress in 2005 to use federal funds to construct a bridge to replace the ferry between Ketchikan, Alaska, a town of 8,900 people, and the Ketchikan International Airport on Gravina Island, population 50, at a cost of $398 million. After it became a national embarrassment, Governor Sarah Palin canceled the project. The same bridge idea reemerged in 2013 with a smaller price tag, but still providing fodder for complaint about high-cost, low-benefit projects.

In contrast, consider the Tunnel to Tomorrow. That's my name for the Port of Miami Tunnel, where the numbers look a great deal better. In a region where more than 175,000 jobs depend on the seaport, serving a nation whose consumers and businesses depend on our ability to move goods into and out of the country by ocean, this public-private partnership is a harbinger of the future for infrastructure finance. This is a $1 billion project to take trucks from the seaport directly to the interstate highway and remove them from congested city streets. The partnership is housed in a project company, called Miami Access Tunnel, set up to operate the concession from the county, which is 90 percent financed and owned by a fund raised by a French infrastructure finance company, Meridiam.

Meridiam fronts the money for the Florida Department of Transportation (FDOT), Miami-Dade County, and the City of Miami in return for performance-based payments over thirty-five years; tax-exempt bonds and federal grants will help ensure that those payments are made, without raising taxes or taking funds from general revenues. This combination has ensured that the construction stuck close to schedule and stayed under budget.

There are numerous reasons we should go deeply into this tunnel. It addresses both current transportation pain points and future economic opportunities—the winning combination for high-priority, nationally significant projects. The pain points include traffic congestion in downtown Miami, which frustrates commuters, causes costly delays for truckers carrying goods, and makes residential development difficult. Miami leader Maxeme Tuchman also found, in her studies while at Harvard, that truckers complain about the gas wasted in idling, since many of them are owner-operators who feel the pain in their wallets. The large economic opportunity involves modernization of the port, an $18 billion economic driver for the region, in anticipation of larger ships as a result of the Panama Canal expansion, enlarging international trade all over the United States. Another opportunity is downtown development on mostly truck-free streets.

The project involves innovation at every point from construction to operations. The French contractor used a new German tunnel-boring machine that did not disturb the ocean floor and thus was environmentally sound, and the tunnel comes with its own digitally equipped operations control center monitoring numerous sensors and safety features—a model for roads of the future. There was strategic thinking about the future of the city, making the tunnel part of a megaplan for an arts-focused, pedestrian-friendly city. Strong committed leadership kept the project alive during numerous crises when it was near collapse. And the project has not required raising taxes or using general tax revenues. Private capital made public commitments possible. The private-sector partner takes on risks that would normally be borne by a public entity, and also engages the public by working closely with the community.

This combination of factors—strategy, innovation, leadership, and partnership between the private and public sectors—makes the project a potential national model. At the same time, the very long approval process over nearly a decade, byzantine federal requirements to tap funds, the vacillation of state and local officials, and the absence of U.S. companies from significant projects like this one illustrate some of the weaknesses in the U.S. transportation and infrastructure financing system. Despite all this, the Miami tunnel became the project that wouldn't die. It proves my adage that everything can look like a failure in the middle; leaders find a way around obstacles and persevere to success. But some obstacles shouldn't be there in the first place.

My point in starting with this story is that questions of finance and political will, of resources and leadership, are intertwined. It is a matter not only of finding the money but also of identifying the costs and benefits that the public will understand and support. These are political, not technical, considerations.

Let's begin with a look at the setting. It helps explain why some leaders found the political will to keep the idea alive amid numerous crises, and how they found the money. It's a thirty-year drama complete with its own spoof movie. Then, with this model in mind, the larger questions of financing the future for America's infrastructure can be addressed.

GOING UNDER, AND COMING OUT THE OTHER END: THE MIAMI PORT TUNNEL AS A NATIONAL MODEL

The Port of Miami (familiarly known as PortMiami) is located on Dodge Island in the sparkling turquoise waters of Biscayne Bay, adorned with glamorous cruise ships and near the desirable southern tip of Miami Beach. It is vital to the regional economy. In 2012, the port's cargo facilities handled 909,197 TEUs (twenty-foot equivalent units, a freight train container), making it the nineteenth-largest U.S. port by overall cargo volume, the ninth-largest container port, and one of Florida's three largest, along with Port Everglades in Fort Lauderdale and the Port of Jacksonville.[1] PortMiami bills itself as the "Cruise Capital of the World," handling 4.1 million cruise passengers in 2011, more than any other American city. The port contributes an estimated $18 billion annually to the local economy and directly and indirectly provides 176,000 jobs, second only to the airport in economic significance for South Florida.

Commercial seaports and waterways are important to all of us— consumers, businesses, and the U.S. economy. They move 2.3 billion tons of cargo a year, many times that of air, including important U.S. exports in agriculture, manufacturing, and energy products. Nearly 80 percent of the volume of U.S. international trade passes through U.S. ports, and waterborne business activities contribute approximately $3.15 trillion to the U.S. economy. In addition, more than 13 million people are employed in port-related jobs. The volume of international trade through ports is expected to double from its 2001 level and the volume of containerized trade to triple. But ports don't sit there by themselves; their value comes from land-based transportation connections. Shippers rely primarily on highways and railroads to transport cargo to inland markets. Congestion or inadequate roads, bridges, tunnels, or railways cause delays and raise costs.[2]

Clogged ports and adjacent congestion jeopardize the everyday low prices Americans have come to expect. Retailers complain about rising costs from port inefficiencies, as I heard from a range of household brand names: former Toys R Us CEO Gerald Storch, BJ's Wholesale Club CEO Laura Sen, DSC Logistics CEO Ann Drake, and the supply chain heads of WalMart and Staples. They cite equipment shortages, too much paperwork, labor unrest, and battles between trucking companies and ocean liners, but keep coming back to port and traffic congestion.

The consequences are felt at the kitchen table. PortMiami brings us fruits and vegetables year-round, such as fresh blueberries from Chile or bananas from Guatemala. It has 1,000 refrigerator plugs, moves 900 refrigerated containers a day, and uses handheld devices to speed Customs and Border Patrol inspection time, processing many containers in minutes. It also carries our agricultural exports; the port moved 31 percent of U.S. waterborne grapefruit exports in 2011. There's more. Imports of apparel, flip-flops from Brazil, clay and cement tiles come in, and exports of packaged foods, electronic equipment, cars, and other manufactured goods go out.

Anticipating high growth after the 2015 opening of the widened Panama Canal, PortMiami director Bill Johnson and his team undertook a $220 million deep-dredging project and the installation of larger gantry cranes to make the port one of only three U.S. Atlantic ports (and the only one south of Virginia) equipped to handle large post-Panamax ships.[3] The cranes tower over their counterparts, looking like massive sculptures on the edge of the water. Larger ships require much deeper channels, even larger cranes, and longer docks, and also have implications for land transportation. Consider this startling statistic: large Panamax ships can carry more than 10,000 twenty-foot containers (TEUs); end to end, that would be nearly a forty-mile-long train. Forty miles' worth of containers? No wonder there are worries about truck traffic to and from the seaport.

Vehicle access is a long-standing problem for PortMiami, as it is for Long Beach and Los Angeles in California. In Miami, cars and trucks have had to travel through the heart of downtown to reach the port bridge. On any given day, as many as 16,000 vehicles—up to 5,000 of them eighteen-wheeler trucks—made this journey, causing numerous significant problems not just for shippers but for everyone else, whether city residents or commuters: congestion in downtown Miami; streets where trucks shut out all other uses; excessive air pollution; and safety concerns.[4] Then there was the kicker for a city becoming focused on arts and entertainment. The port entrance and exit is near the Arsht Performing Arts Center and the American Airlines Arena, where the popular Miami Heat play basketball; game nights produce gridlock nightmares.

What to do about all of this? The idea of an access tunnel had been a dream since the 1980s. At least, a dream of some people. It was also highly politicized, as infrastructure shouldn't be but often is. Let's look inside several rounds of a long drama, in which politics had to be overcome and money found. Four rounds of politics are interspersed with three rounds of financing dilemmas. By observing the characters in this seven-act drama, we learn exactly how a major infrastructure project unfolds or is delayed, and all the players who must get involved to find the money. It's a fascinating story of persistence and creative leadership, on the way to achieving quintuple wins.

POLITICS ROUND 1: THE VISIONARY ENGINEER

Jose Abreu is the first hero of this story, widely credited as being an early visionary and the champion who persevered.

Jose Abreu emigrated from Cuba by himself at age thirteen and graduated with a civil engineering degree from the University of Miami in 1976. He has worked in the Miami area for the greater part of his life, mostly with the Florida Department of Transportation. A series of promotions led to his appointment as secretary of transportation under Governor Jeb Bush in 2003, a position he held until 2005. In 2005, he left FDOT to direct the Miami-Dade Aviation Department, which owns Miami International Airport (MIA) and five general-aviation airports. During his eight years there, he brought costs and processes under control, undertook ambitious development plans, and improved the airport's credit rating— another sign of his talent. In 2013, Abreu took an executive post with Gannett Fleming, as infrastructure firm, still based in Miami. His life after FDOT is relevant, because he continued to keep his eye on the tunnel even when he moved from seaport to airport.

In 1982, Abreu, then a young engineer in his mid-twenties, was working for the Miami-Dade seaport. Abreu realized that "you can't survive as a seaport without being connected to the expressway." In Miami, the interstate constitutes only 3 percent of the lane-miles but carries 70 percent of all truck traffic. An old mechanical drawbridge to Dodge Island was inefficient and caused enormous traffic congestion where it met downtown Miami. The Federal Highway Administration and the State of Forida were unwilling to help pay for a new bridge until Miami could prove the bridge's long-term viability.

Abreu and his colleagues started investigating other ideas for bridges or alternatives. Abreu was assigned to lead a dozen engineers to design a possible tunnel between the port and the interstate. When Abreu presented his team's work on a tunnel, the head engineer at the port showed up right at the end of the presentation. He asked whether he had missed anything, and someone who had seen the whole thing told him, "Nah," Abreu recalled. At that point, the tunnel idea was more or less dead.

In early 1984, Abreu joined the FDOT, overseeing an office that managed federal grants. Then–Congressman Claude Pepper had negotiated a $4 million federal appropriation to fund further study of a tunnel at the Port of Miami, and this grant came through Abreu's office. Looking back, he jokes, "The tunnel was following me." As studies continued, Abreu became convinced that a tunnel was the only answer. City of Miami mayor Maurice Ferre, who oversaw the new city skyline, felt the same way. However, tunnel-boring technology was still unproven—this was well before the Chunnel under the English Channel—and other methods of construction were environmentally problematic. Moreover, by that time money for a new port bridge had been received. Despite Abreu's own conviction that the port needed a tunnel, the idea was still unpopular. Many years would pass before others bought in—and before Abreu was in a position to make the project a reality.

In 2000, federal transportation officials approved the concept of a tunnel. Nothing much happened until 2003, when Abreu was appointed Florida Governor Bush's secretary of transportation. By then, he knew the tunnel was technically feasible; the question was how to move on it. He presented the idea to Bush, who said, "Prove the engineering validity." That, Abreu says, took about two minutes because he was "super prepared." Bush also wanted to make sure the business community was on board. Abreu had reached out beforehand to business leaders, and he more or less told Bush, "They're waiting in my conference room for your go-ahead."

A few months later, Governor Bush—whom Abreu calls "extremely thorough"—had also met with business leaders and started to talk specifically about funding. Bush, who was a South Florida resident, insisted on getting involvement from the City of Miami and Miami-Dade County and wanted to make sure there was federal money as well. Abreu worked to get buy-in from all parties. In 2005, in the waning months of the Bush administration, Abreu resigned to become director of aviation for Miami-Dade County two weeks after the county took over the beleaguered MIA North Terminal project from American Airlines, successfully completing a project that was plagued with cost overruns and lawsuits. He even lowered the fees airlines pay to land at MIA and increased the airport's credit rating. Abreu earned respect as a great manager for good reasons.

Despite his new focus on the airport, Abreu remained in the tunnel loop because of this respect and his wealth of knowledge. For example, a meeting between his FDOT successor Denver Stutler and Miami-Dade County mayor Carlos Alvarez took place in a MIA conference room so Abreu could join. When the tunnel project ran into resistance from cruise lines, which were worried about a possible cave-in, the coalition in support of the tunnel overcame these objections; Abreu was involved in an unofficial, but very influential, capacity. "My contribution was peppered into the conversation," he said later.

MONEY ROUND 1: A PRIVATE-SECTOR FINANCIAL PARTNER

The project's complexity made Kevin Lynskey, a key port official, think at times that it was "a massive financial engineering project with a little construction on the side."

FDOT officials sought a way to spread the risks of such a large, costly undertaking, so that they would not have to shoulder all of it themselves. Someone suggested an availability payment structure that would involve granting a concession to a private-sector funder and operator. The language of public-private partnerships—now familiarly known as PPPs or P3s—was still new.

In 2005, FDOT performed a "public sector comparator" that estimated the costs of delivering a tunnel by traditional methods at $1.2 billion. FDOT hired Jeffrey Parker & Associates, a PPP advisory firm, and worked with the City of Miami and Miami-Dade County (the port's owner) to determine how to deliver the project. The first funding model discussed was shadow tolls (per vehicle payments from the government to the private operator). Early on, project sponsors had decided they did not want to toll drivers directly. At the time, that was the traditional method for public-private partnerships to enable the private investor to earn a return. But shadow tolls would have exposed the private partner to the same demand risk as traditional tolling, that is, whether there would be enough paying users. FDOT was willing to absorb this risk, so the team decided it would employ an availability payment structure instead, whereby the winning bidder would receive payments for reaching construction milestones and then receive annual payments for operation and maintenance of the facility. In 2006, FDOT issued an initial request for proposals from private partners, inviting them to beat the public-sector comparator estimate. Bidders would have to maintain design standards and demonstrate their ability to finance the project and shoulder construction risk.[5] Payments would start only after the work was done.

The winning project company, Miami Access Tunnel (MAT), began as a joint venture led by Bouygues Civil Works Florida, a division of Bouygues Travaux Publique, a French construction concern with extensive tunnel-boring experience that included work on the Channel Tunnel between England and France. Bouygues submitted a design that used innovative drilling techniques to avoid blasting the ocean floor. A cutting-edge tunnel-boring machine would be imported from Germany to drill two 4,200-foot-long tunnels, which would be sealed to prevent flooding during hurricanes with massive metal gates. Innovative freezing and caulking techniques would be used to harden the ground and ensure that the tunnel remained watertight. All this, Bouygues estimated, could be done for only $610 million—half the expected public cost. Then, after thirty-three years of construction and operation, MAT would hand the tunnel back to the state in first-class condition—or pay a hefty penalty. Not surprisingly, this sounded good to FDOT and would be good for the public too. Bouygues won the contract.

POLITICS ROUND 2: THE MAYOR AND THE MEGAPLAN

Federal and state officials didn't want to go ahead unless there was 100 percent local participation. Although the State of Florida had agreed to pay most of the project costs, Miami-Dade County and the City of Miami were required to approve their share of spending by a drop-dead date in late 2007, after which the state had threatened to abandon the project. The city was required to commit $50 million that would be paid out of Community Redevelopment Agency funds, avoiding the need to dip into general tax revenues. Despite the relatively small commitment required, and the fact that in 2004 voters had actually approved $100 million for the tunnel, this required a vote of the city commission (Miami's name for a city council).[6] Without local matching funds, federal funds would be lost, and the project would have to start again at the back of the line.

Then–City of Miami mayor Manny Diaz knew that the state and county were unhappy with the city because the city was the only party that had not approved its funding responsibility yet. While having initially been enamored of a signature bridge like Boston's Zakim Bunker Hill Bridge, Diaz supported the tunnel in concept because the tunnel would take trucks out of his urban core. The tunnel and light rail were central to his vision for enhancing quality of life in the central city. He wanted the Miami waterfront to be a pedestrian-friendly district for performing arts, museums, parks, restaurants, nightlife, and residents in new condo developments. But he knew that the tunnel project was still controversial, and he knew more than the planners seemed to about local politics. Infrastructure projects don't proceed just because they're technically feasible and funded.

Diaz, who served from 2001 to 2009 and couldn't run again because of term limits, was a Cuban American lawyer who had never previously run for office but stood out as a leader in the region for being untouched by scandal. He stabilized city finances, inspired voters with visions for Miami's future, won awards for his service, and enjoyed sufficient respect among his peers to become head of the U.S. Conference of Mayors.

Diaz recalled a particularly contentious meeting in the late summer of 2007 with representatives from the FDOT, who were pressuring him to get the tunnel project to a vote of the city commissioners. "There were all kinds of deadlines," he recalls their saying. "If not met, they'd lose money, and the project would go away. They were somewhat nasty. There was heavy-handed pressure. 'You've got to do this,' they said. I said, 'Look, the votes are not there; only one member of council has expressed support.' September was the end of [the] fiscal year, and we were talking about budget cutting. In the middle of budget debates, to spend $50 million, even if from a different budget? It's the optics. This thing will go down in flames.

"I said, 'You've got to give me more time. I'm not taking this to commission to get blown up.' I was absolutely sure it would fail. I held my ground and said absolutely not, I'm not taking it to the council. They said we're going to lose it. I said you'll lose it anyway if we take it to a vote. I could have easily said 'OK, fine, I'll get a vote,' but it wouldn't pass."

Despite strong arguments for the tunnel, such as competition with Fort Lauderdale and Jacksonville for port business (and its previous popular approval), there were side issues. Bouygues had affiliates doing business in Cuba, where Dubai was financing a $300 million port construction in Mariel, Cuba, the place from which a major wave of Cuban emigrants left for Miami. Some segments of Miami's activist Cuban exile community vociferously opposed Bouygues for supporting Cuba (and competing with Miami for international shipping), and that noise had to be quieted.

Diaz feels he had a good relationship with the commissioners. In his eight years in office, he never used his veto. That wasn't because the commission was a rubber stamp for the mayor, as some critics argued, but because, Diaz says, they were constantly meeting, sitting in a room together for hours at a time, constantly working through differences. By the time a vote was held, adjustments had been made, and the commission was in sync. That's one reason Diaz could withstand FDOT pressure.

With no time to lose and a vital project on the line, Diaz pursued a creative but controversial strategy. He bundled a number of pending projects favored by various commissioners—a new baseball stadium, a streetcar system, a performing arts center, and a museum park— into a $3 billion umbrella "megaplan" to revitalize Miami's urban core. He asked for a vote on the entire package, even though each project still required a separate vote. Although these projects had all received public approval on their own, each had gotten tangled up in political hurdles in the time since. Diaz's audacious plan worked: commissioners realized that they would need to support each other's projects to get their own projects endorsed, and the megaplan was approved by a 4–1 vote. Then each particular project was voted on individually. The tunnel passed by a 3–2 vote.

Critics of the megaplan alleged in court that it amounted to closed-door wheeling and dealing, but the Florida Supreme Court sided with the city and preserved the deal. The project could move forward. In fact, even before the concession agreement could be signed, the project was recognized as the 2007 public-private partnership project of the year by the American Road & Transportation Builders Association. (But an infrastructure finance expert downplayed the award's significance, saying that awards have been given to projects that subsequently failed.)

MONEY ROUND 2: THE FINANCIAL CRASH

All looked good. In fact, very good. And then the global financial crisis intervened. Bouygues had initially partnered with a consortium of financial firms led by Lehman Brothers and planned to issue hundreds of millions of dollars in private activity bonds (a form of municipal bond supporting a private-sector development) to finance construction. Babcock & Brown, an Australian infrastructure investor, would provide equity along with Bouygues.

Lehman Brothers went bankrupt in mid-September 2008, removing the lead underwriter. As the global financial crisis erupted, Babcock & Brown also imploded; its market cap went from over $8.5 billion in mid-2008 to $50 million in almost no time—in part because it was overleveraged and undercapitalized, which seemed to characterize many of the infrastructure funds at the time, which did some of the same risky things as real estate speculators. Babcock & Brown soon collapsed too.

The failures left the Miami Access Tunnel company and its public partners without a lead underwriter or a majority equity partner. Spooked investors were driving up rates for private activity bonds, which I'll describe later in this chapter, and they would no longer be a cheap enough source of financing. Jeffrey Parker, FDOT's financial adviser, stepped in with a plan to find another equity partner, promising that the firm could do so without tainting the transparency and fairness of the government's initial bidding process. Paris-based Meridiam Infrastructure Fund, a recently formed infrastructure investor whose North American operation is chaired by Jane Garvey, FAA administrator in the Clinton administration who also served on President Obama's transportation transition team, and Joseph Aiello, an experienced project financier, had admired the project from afar. Meridiam had been organized too late to take part in the initial request for proposals, but at the time of the meltdown, Meridiam had already begun conducting due diligence in anticipation of a small equity stake. When Babcock & Brown requested that Meridiam take over its equity, Garvey and Aiello leapt at the opportunity.

Meridiam, by the way, took a more cautious approach to infrastructure investment than some other companies, had a long-term time horizon and didn't churn assets, and maintained a close connection to project management, all of which made it more seaworthy in the wake of the financial crisis.

Despite this fast action, Florida secretary of transportation Stephanie Kopelousos under the next governor, Charles Crist, who lacked Jeb Bush's ties to South Florida, announced on December 12, 2008, that the project would be canceled. Some observers suspected it was politics—bad blood between administrations. FDOT leaders told project sponsors that they weren't trying to kill the project outright, but they wanted a new RFP (request for proposal), hoping to get even more favorable terms for the state—in essence starting all over again. The project had been so long in the making that delay was tantamount to a death sentence.

POLITICS ROUND 3: SAVING THE TUNNEL IN THE STATEHOUSE

With Meridiam lined up, the tunnel project's advocates launched a "full-court press" in the state Capitol, lobbying the governor to reverse course. Meridiam now had the burden of proving that the project was finance worthy; it worked on cost reductions, such as getting Bouygues to slash its estimates and accepting a cut in its own rate of return, making it harder for FDOT to turn down its proposal. There were behind-the-scenes conversations and dramatic public events. Then Miami-Dade County mayor Carlos Alvarez announced at a 10 a.m. press conference on a day shortly after the FDOT announcement that he wasn't taking no for an answer, either, and was sending a team, including Jose Abreu, to Tallahassee to negotiate. Shortly after the press briefing ended, he contacted Abreu and told him to pack his things—he was going to Tallahassee at 5 p.m. the same day with Bill Johnson, the Port of Miami director, Kevin Lynskey, now deputy director, and others.

"I was up there for two days," Abreu recalls. "The first day had everything to do with the procurement process. Basically, while the contractor was the same, the finance partner had changed and thus by definition the FDOT said the concessionaire had changed, and procurement needed to restart. We argued that the plans, specs, and construction cost were the same, and as long as the interest rate was the same, what is the problem with a substitution? Attorneys went back and forth, legislators weighed in, and by the end of the day we all agreed that we would try one more time, just one more time!

"Day two, the group was much, much smaller. Three of us from the county: County Manager George Burgess; Kevin Lynskey, a financial guy who now works as a deputy director at the Port; and me. Three folks from DOT as well: two assistant secretaries and Gus Pego, the district secretary, who was crucial. The challenge was to present a 'take or leave it' document so detailed that there would be no room for anyone to protest a subsequent award," Abreu says. Lynskey also recalls working with a few favorable legislators who could hold up the state budget while this was resolved.

After tense negotiations and a private bus tour in Miami with state lawmakers and local leaders on January 18, 2009, Stephanie Kopelousos said, "We are taking a step back and looking at what options are available to us." Shortly thereafter, FDOT agreed to continue support for the project and allow Meridiam to replace Babcock & Brown. Abreu observed later, "It looks like we got it right."

MONEY ROUND 3: A BETTER PARTNER, A BETTER DEAL

Kevin Lynskey recalls that he was one of about eight people in the room when three bidders were whittled down to Meridiam. After signing on, Meridiam searched for a new investment plan, and by October 2009 had succeeded in securing $341.5 million in low-cost senior bank loans from a consortium of mostly European banks. Milestone and acceptance payments from FDOT to the Miami Access Tunnel company of $450 million would pay off the majority of project company MAT's bank debt in 2014, while the first of thirty annual $32.5 million availability payments would pay off the remainder.

The terms of the deal were more favorable to the public than the Babcock & Brown deal, and a new twist was added. Meridiam and its partners secured a $341 million federal TIFIA loan. TIFIA (Transportation Infrastructure Finance and Innovation Act) was approved by Congress in 1998 to provide loans, loan guarantees, and lines of credit to "leverage limited Federal resources and stimulate private capital investment" in infrastructure projects with national or regional significance.[7] The TIFIA loan was another reason Meridiam slashed its rate of return; to be fast-tracked and jump ahead in the queue, the project would have to offer more money up front. The loan replaced part of the proposed contributions from the county and the city. Meridiam would repay the loan from the availability payments for subsequent years of the thirty-year operating concession. The federal contribution was a good financing source since other options, such as municipal bonds, were off the table, as Lynskey explained. "It was the financial crisis. Private activity bonds wouldn't fly"— because the private activity bond market had practically evaporated, and it would be hard to raise money in that context.

To top it off, Meridiam and Bouygues contributed $80.3 million in equity financing, split 90/10, and secured a $180 million contingency fund in the event of geotechnical complications out of MAT's control.[8] Risk management was a good thing for a project that promised not to disturb the ocean floor.

The contract structure was an impetus for the partners to deliver the project efficiently; MAT Concessionaire would get paid only upon completion of specific milestones, with payments deducted for failure to meet performance objectives or deadlines. Upon final completion, the project transitions to a thirty-year operating concession with annual $32.5 million availability payments from FDOT to MAT Concessionaire for service and a stipulation that the tunnel be transferred back to public hands in "first class condition" at the end of the concession period thirty years later. The maintenance stipulations also provide an incentive for high-quality construction work; shoddy construction would raise future maintenance costs and eat into

MAT's profits, as would lapsed maintenance, lane closures, and other deviations from predetermined maintenance standards. As a result, some at Meridiam like to say they "live in the world of no excuses."

POLITICS ROUND 4: COMMUNITY COALITIONS

Christopher Hodgkins, a former Massachusetts state legislator working in Florida, approached Meridiam after recognizing Garvey's and Aiello's names in the news. During a breakfast meeting in Miami, they became convinced that they needed someone like him to manage the complexity of local community relationships during and after construction. Hodgkins joined MAT as vice president and the company's public voice.

Jobs are often a rationale for infrastructure investments; U.S. Chamber of Commerce and AFL-CIO presidents Thomas Donahue and Richard Trumka joined hands after the financial crash to champion infrastructure as a source of job creation. In Miami, the jobs angle was a help in getting community buy-in. The construction phase would employ more than 6,000 people, 82 percent of them from the region. MAT hosted employment fairs as well as vendor and supplier outreach. One major push was "Operation '305,'" named for the local telephone area code; it included a supplier and contractor expo for local companies to learn about available work, how to apply, and how to win contracts. MAT also reached out to the Spanish-speaking community. Community partnerships included numerous chambers of commerce—Greater Miami, Miami-Dade, Miami Beach, South Florida Hispanic, Haitian American, and French American—as well as the business-oriented Beacon Council, World Trade Center Miami, Miami-Dade County Public Schools, and two unions—the International Brotherhood of Electric Workers Local 349, and the Laborers' International Union of North America. Quite a visible jobs program.

After a period of design and preparation, MAT was ready to start building. The $45 million tunnel-boring machine arrived in late June 2011 by cargo ship. It was a lot of cargo: 3,000 tons in weight, 42.3

feet in diameter, 428.5 feet in length. A heartwarming community effort, a naming contest with local Girl Scouts, brought the machine down to human proportions. The Girl Scouts' winning name was Harriet, after African American abolitionist Harriet Tubman of Underground Railroad fame. ("One of the few women in infrastructure," Maxeme Tuchman exclaims, "turns out to be a machine.") From November 2011 to the end of July 2012, Harriet dug the westbound two-lane segment, then turned around to dig the eastbound two-lane segment, finishing in May 2013. The lowest point was 120 feet below sea level.

Installations within the tunnel followed, high-tech and state-of-the-art. Less than a mile in length in each direction, the tunnel included numerous redundant safety features. There are 44 jet fans, 91 cameras, 42 emergency phones, 50 heavy floodgates, 5 cross-passages between the two tunnels, and an extensive system of infrared sensors, all connected to a surface operations control center. MAT's Hodgkins calls it the safest tunnel in the world.

A new lane was added to I-395, McArthur Causeway to Miami Beach, with an exit for the tunnel that would eventually handle trucks and buses carrying passengers to cruise ships. Remaining traffic congestion on the causeway had little to do with the tunnel construction and everything to do with the failure of Miami Beach to embrace Miami mayor Diaz's 2004 proposal for light rail between the City and the Beach—a proposal that has recently reemerged out of necessity.

Sure enough, geotechnical complications created unexpected tunneling costs and necessitated dipping into $64 million of the contingency fund. Still, project managers boasted in December 2013 that the project was "on time and on budget"—a rare occurrence in one-of-a-kind transportation projects. Teams worked around the clock to finish by the May 2014 deadline for the full $350 million acceptance payment.[9]

In March 2014, Manny Diaz and I toured the almost completed tunnel with Chris Hodgkins, who drove us in a four-person Caterpillar construction vehicle. We met in the construction office trailer for the mandatory hard hats, fluorescent safety vests, boots, goggles, and PowerPoint information briefing. Inside the tunnel, workers were fitting an extra layer of panels over the usual tunnel interior. The sections that were done had a sleek, modern look, as did the overhead lights and fans. One of the coolest features, I thought, was the warning system for over-height trucks: bells and whistles, clanging items that drop onto the truck roof, and if all that isn't enough to stop a truck, gates that automatically close, thanks to the sensors and control center. I continue to consider this a model for the future of roads.

If ever a construction project could appear warm and fuzzy, this one did. Harriet was long gone, but her size was apparent from the massive half-round entrance we saw; she was often invoked to personify the excitement. Pushing national hot buttons such as STEM education, MAT sponsored an educational outreach school program for National Engineering Week and a tunnel exhibit and information kiosk at the neighboring Miami Children's Museum featuring Bob the Builder. Hodgkins and his team from Meridiam are masters at community relations; still, it took extra convincing to get the French parent to understand why money was spent on non-construction-related activities—and compared with other foreign firms, Meridiam is considered enlightened in this regard.

A case in point: Manny Diaz and I watched "Going Under," an entertaining just-produced video presenting a tunnel-opening preview as a spoof movie preview. First a "ratings" screen: "MT: All Audiences," "For Everyone Who Ever Dreamed of a Miami Tunnel." MAT Concessionaire and Miami-Dade County are the "studios" for "A Miami Tunnel Production," with FDOT and PortMiami as executive producers. The drama begins with Jorge Mora, a driver for Southern Cartage, exclaiming, "Oh, no, look at all this traffic!," as well as some pedestrians disclaiming the sorry state of Miami's truck-filled downtown. There follows footage of Miami leaders—ending with port director Bill Johnson—expounding on the need to build a port tunnel. Then comes the "Going Under" title screen, an underwater animation. Next are shots of delighted pedestrians in downtown Miami, and then workers headlining a "cast of 968" with "6,000 extras" and starring, naturally, Harriet. The video returns to truck driver Mora, who expresses his delight in the new tunnel. Then, in a capstone moment, Jose Abreu stands in the nearly finished tunnel and says, "Billy [Bill Johnson], I told you a tunnel would work!"

"Under budget" was achieved, although "on time" slipped a little bit. As a result of a few maintenance tweaks, such as redoing all the mountings for exhaust fans because higher-than-expected vibrations knocked out two of them, the actual tunnel opening was delayed until late July. Bouygues paid $115,000 a day in fines for every day of delay. Some suspected that the county was acting too slowly on inspections and certification, such as the certificate of occupancy for the tunnel operations center, but that's just speculation.

The ribbon-cutting ceremony proceeded as scheduled on May 19, 2014, replete with federal, state, and local officials. A bus carrying VIPs to the ceremony, driven by Eduardo Basulto, became the first nonconstruction vehicle to traverse the tunnel. Upon emerging on Dodge Island for the ceremony, the bus riders erupted into cheers. Workers paraded through the tunnel accompanied by a costumed band and a tractor-trailer—the first truck to make the trip, though the tunnel wasn't yet cleared for traffic. Governor Rick Scott led the ceremonies. U.S. Secretary of Transportation Anthony Foxx hailed the project as "good for America."[10]

Jose Abreu was on the bus. He was recognized for his decades of work for the tunnel. He is very proud of it—a "marvel of engineering" and the driest tunnel in the world. He is quick to point out that credit really goes to Bouygues for actually getting it done. He told us that the feeling for him was "like giving birth after thirty years."

Manny Diaz was there and reported that he felt great. "It is obviously a project that everyone embraces now. At the opening everyone was there, including a former city commissioner who voted against it every time it came up." That former commissioner is Tomas Regalado, Diaz's successor as mayor. He's now proud of his tunnel, which officially opened to the public in early August 2014.

Once the novelty and the fuss are over, the tunnel will fade into the background as just another piece of taken-for-granted infrastructure. But every time Maxeme Tuchman rides her DecoBikes bicycle over the causeway into the truck-free streets of downtown Miami, she and thousands of peers will benefit from the new zones open for pedestrians, cyclists, arts audiences, restaurant-goers, and new homes, surrounded by cleaner air. Not to mention having fresh bananas year-round at an affordable price coming through the bustling port.

If this can be done in Miami, why not everywhere? The dynamics are echoed in every community and region in America. The construction is fast. The politics are slow. The money parts require an openness to collaboration and the development of a public-private partnership in which the main private-sector partner knows how to keep the public interest in mind. Federal and state funds are critical, and so is leadership by the county and city. A cross-sector, multistakeholder coalition, with shifting members, propels the project. Leadership overcomes numerous obstacles to create the quintuple wins: safety, convenience, productivity, sustainability, and economic opportunity.

Let's take a step back to see how and why it's so difficult to get infrastructure projects off the ground (or below the ocean) in America, and what new approaches should be considered.

BONDED: CLASSIC INVESTMENTS AND THEIR YIELD FOR AMERICA

America has a long, strong tradition of financing long-term infrastructure at the regional level through municipal bonds. This system has served the nation well, and will continue to do so, but it is no longer enough.

I'm personally grateful for municipal bonds. I went to college on municipal bonds. When I was a little kid in a middle-class family, my father clipped coupons, and I sometimes watched. It was a simpler time. There was a large middle class living the American dream, college was more affordable, and everything was still on paper. "Munis"—tax-exempt bearer bonds—were my dad's favorite form of saving. He wasn't trying to make a killing, just collect steady interest and use it to buy more bonds. He had a stash stored in a bank safe-deposit box, with semiannual payment coupons attached, special scissors to cut them, and windowed envelopes for depositing them to his bank account. My family was paying for bridges, tunnels, turnpikes, ports, industrial parks, hospitals, and school construction. Although my dad wouldn't have called himself a source of "infrastructure finance," he was. Cities, counties, and regional authorities offered him tax-free interest payments, coupon by coupon, until the bonds were redeemed many decades later. It was safe and predictable.

Back then, local governments could, and today still can, also issue bonds on behalf of private business entities, giving some businesses the benefit of low-cost financing. In 2005, the federal surface transportation act added highways and freight transfer facilities as eligible projects for these "private activity bonds" and authorized up to $15 billion for transportation projects. As of 2013, only about $4.5 billion in transportation private activity bonds were issued, with another $5.2 billion approved, barely registering as a source of infrastructure finance. But groups outside the United States think this is a good deal. For example, the State of Indiana, helped by Citigroup, issued private activity bonds in July 2014 to I-69 Development Partners, owned by Isolux Infrastructure Netherlands and Canada's main public-sector pension board, to design, build, maintain, and operate a twenty-one-mile section of Interstate 69.[11] Can we rebuild America twenty-one miles at a time?

Those years of my dad's coupon clipping coincided with the glory days of American infrastructure development. Municipal bonds remain a popular form of finance for regions, and they were part of the financing mix for the Port of Miami Tunnel. But this traditional vehicle won't be nearly enough. We need more than ad hoc stretches of highway.

The American Society of Civil Engineers estimates that the United States will require investments of $1.98 trillion by 2020—compared with an anticipated $910 billion in available funding—to maintain and upgrade the nation's decades-old highways, inland waterways, ports, airports, and rail and transit systems, which does not include the vehicles that operate in these systems and the communications infrastructure connecting them.[12] The National Surface Transportation Policy and Revenue Study Commission, convened in 2005 to investigate the scope of transportation funding challenges, argued that there's a need for at least $225 billion annually for surface transportation alone through 2050.[13] Yet the share of American GDP devoted to infrastructure investment is only 2.4 percent, compared with an average of 5 percent in European Union countries and as much as 9 percent in China and other emerging economies.[14] America must do better.

As I write, the Highway Trust Fund is running out of money. It might be saved, but it has been running out of gas (apologies for the pun) because of an obsolete premise: a federal gasoline tax to pay for roads, with a small portion for public transit. Other federal funds are down and dying. User fees are on the rise, aided by technology. But passenger ticket fees are not enough to modernize airports or bring us NextGen air traffic control, and there's a limit to what the airlines can charge.

It's time to rethink how the U.S. system operates. The first question—and one of the first that Florida DOT officials asked—is what role the private sector should play.

SELL! THE SEDUCTIVE (BUT OFTEN WRONG) DREAM OF GETTING CASH FOR EXISTING ASSETS

Hey, here's an idea. Maybe the public sector could get out of the infrastructure business altogether by finding private owners for public assets and raise cash in the process? Not a good idea. The problem is that these efforts have largely failed.

For many years, governments have been trying to find ways to get the private sector to pay for public infrastructure projects. In some cases, cash-strapped state and local governments, whose voters are unwilling to entertain tax increases, have looked for what they own that they could sell—or lease on a long-term basis. The federal government took this tack, too, with the FAA's program to sell airports to private investors, hoping they could bear the costs of modernization and run them more efficiently. But as I indicated earlier, private investors are not buying airports.

By privatizing and monetizing existing public assets, public authorities can tap deep-pocketed private investors who could provide an enormous capital influx for much-needed development, proponents hope. They argue that injecting a profit motive into infrastructure management is necessary to reap efficiency gains and save money, and that monetizing assets via privatization is necessary to cover budget gaps and build for the future.

But that's only one side of a controversy. Others have derided the desire to monetize assets as "selling the family silver" for short-term goals and suggested that privatization constitutes an intergenerational transfer of wealth from future citizens who will have to pay more to use assets that were previously public goods.[15] Stephen Koch, deputy mayor of Chicago under Rahm Emanuel, is on that side. Previously a banker with Credit Suisse, Koch has seen the situation from the private investor perspective too. When I talked with him at his unpretentious office in City Hall after visiting a variety of Chicago projects, he was adamant that privatization isn't a solution. "Is our salvation in private investment in public good projects? Honestly, it's a rare occasion that you will get a better outcome for the public by using private capital over public capital in the U.S.," he declared. "We need to change the national dialogue. We need to get voters to believe that the public sector needs to pay for public infrastructure. We are not better off with private investment; we will pay more. . . . We're running out of living off the last generation's investment."

Chicago's experience could be a teacher for the nation. Privatization has crashed both on high-speed roads and on city streets.

In 2004, a consortium of investors led by Macquarie and Cintra, large infrastructure investors from Australia and Spain, purchased a ninety-nine-year lease of the Chicago Skyway, a 7.8-mile toll road connecting Indiana commuters to Chicago's downtown, for $1.8 billion. The deal marked the first toll road privatization in the United States. A Chicago alderman called the deal "the greatest single financial coup in the history of Chicago"—the winning bid was almost $1 billion more than the other two competing bids—but others claimed that the city was selling the future to cover the costs of poor management in the present.[16] On the day the private operators took over, tolls on the road increased by 25 percent, confirming many citizens' fears that the privatization was simply a wealth transfer from Chicago drivers to foreign investors.

Later commentators concluded that "under all reasonable demand scenarios" the winning bid was too high, constituting a "winner's curse" and guaranteeing that the concessionaire would have to continue raising tolls to cover its interest obligations.[17] Soon thereafter, in 2006, the State of Indiana leased the 156-mile Indiana Toll Road, which connects to the Skyway, to the same consortium for an additional $3.8 billion, seventy-five-year concession—almost $2 billion more than a state-commissioned analysis that put the values of future cash flows at $1.9 billion—prompting a similar outcry from privatization skeptics.

The big losers in the Skyway and Indiana Toll Road deals were investors in Macquarie's funds, which one investor was quoted as calling "the most efficient method of legally relieving investors of their money ever conceived." The structure of the consortia meant that infrastructure assets like the Skyway were owned by shareholders in Cintra and Macquarie's funds, not by the firms themselves, and that Macquarie was paid management fees based on the size of its deals rather than their profits—an incentive to overpay for assets and load operating companies with debt. And this meant that management always got paid, regardless of revenues or profits, which is the kind of financial structure that removes incentives for high performance and often causes trouble.

The Skyway and Indiana Toll Road careened into financial distress. The toll road's losses in 2010 exceeded $260 million, and in 2013 the concessionaire hired restructuring advisers as it struggled to meet a December debt payment. Although the concession makes an operating profit, its debt load climbed past $4.4 billion, and insufficient toll revenues created a risk of default when $3.9 billion in debt matures in June 2015; under the contract, that could allow Indiana to take back the road while keeping its lump-sum payment. Some 75 percent of the Skyway's operating income goes to debt payments, suggesting that future problems are possible. In other North American Macquarie-led toll road projects, investors have filed a lawsuit accusing Macquarie of using inflated traffic projections in its bond pitches.[18]

As if that weren't enough, next came the parking meter privatization debacle. In 2008 Mayor Richard Daley raised cash by leasing the city's parking meters for $1.15 billion for seventy-five years, rushing the proposal through the approval process at a potential cost of billions of dollars in forgone revenue. The deal was almost universally unpopular. Nearly all of the up-front payment was used to cover short-term budget holes, another example of selling the future. The private company, Chicago Parking Meters LLC, quadrupled parking rates and bungled early implementation, such as credit card functionality. Then the company began billing the city for its lost revenue from street repairs, festivals, and disabled parking—charges that could add up to $1 billion over the life of the contract.[19] CPM's 2012 revenue totaled almost $140 million—up 30 percent over the preceding year and five times the approximately $24 million Chicago collected in the year before privatization.[20]

These skewed outcomes do not surprise some leading public financiers who are skeptical about selling public assets to fill public coffers. In 2013, though legally bound to the parking meter deal, Mayor Rahm Emanuel's administration, which included Stephen

Koch, negotiated modifications, such as extending metered hours in the downtown in return for free neighborhood parking on Sundays. Clarifying terms saved the city millions of dollars.[21]

The problem with privatization isn't only whether government should sell; it's also whether anyone wants to buy. In the United States, private investors don't want the risk of outright ownership of existing public assets, as the sorry history of the FAA's program to privatize airports has shown; in twenty years, there have been numerous failures to attract interest and only one operating model before the recent Puerto Rico deal (for which Highstar Capital has high hopes despite a new governor's opposition, which he has promised to put aside). Private investors see politicians as being, well, flaky. In 2007–08, Governor Edward Rendell of Pennsylvania nearly succeeded in privatizing the Pennsylvania Turnpike, only to have the state legislature cancel the transaction at the last moment. In short, there's a long list of pitfalls of privatization, including an opaque public decision-making process, flawed forecasts (a study of 104 new toll roads found that traffic forecasts averaged 20–30 percent higher than reality),[22] and, in many cases that have not been discussed here, poor oversight and management.

CAPITAL PARTNERSHIPS:
THE BEST OF BOTH WORLDS

Such bungled deals notwithstanding, there are still instances where private-sector businesses own, operate, and maintain infrastructure. After all, airline companies run air transportation and construct their own terminals at airports. Investor-owned electric utilities were once municipal power plants. Freight rail companies own and maintain their own tracks, and the system they operate in the United States is thought to be the best in the world.

So we don't want to take the private sector and its competition-honed capabilities out of the picture. But we don't want the public sector out of it either. What if you could marry private-sector effi-

ciencies with public-sector values and protection of the public interest? Combine private investors and public funders to tackle projects
too large for either to handle alone? Align everyone's interests?

Public-private partnerships, PPPs or P3s, arouse growing interest, and could be just what America needs to renew and reinvent its
infrastructure. Freight railroads are strictly profit-making entities,
but, as we have seen, they also collaborate with federal and state governments—and with one other—to finance upgrades of infrastructure
and oversee the process. Partnering with public authorities enables
companies to deliver projects and maintain transportation systems,
and do it cost-effectively, without turning to outright privatization.

That's what makes the Port of Miami Tunnel project a model.
MAT has thirty-year rights but must return the tunnel to the county
in perfect condition after this time, and it gets paid only when meeting certain milestones. If there are shortfalls, the private operators
are responsible; Bouygues paid Miami-Dade County a daily fee for
the modest delays in tunnel completion; any future service failures
will lead to reduced payments to MAT. Advocates argue that by
injecting profit-driven discipline into infrastructure delivery, PPPs
could help correct failures in government provision of infrastructure like slow decision-making, inefficiency, and lack of competition
while also preventing the inequality of access that a purely private
approach could cause.[23] PPPs offer an appealing middle way: delivering public goods at low cost while also providing attractive opportunities for private investment.

The PPP idea is a big definitional tent that holds many different
types of projects. They already span the country, from the I-595 managed lanes in Florida to the Presidio Parkway in California. Although
road projects make up the majority of transportation PPPs in the
United States, the model has begun to make inroads into other transportation sectors like airports, port operations, and public transit;
Denver's Eagle, an ongoing commuter rail project, and Maryland's
Purple Line commuter rail have or are soliciting private partners.

As the Miami case illustrates, these are complex deals. To succeed,

they require transparency, expertise, and accountability on the part
of both public and private partners. That's why the suggestion has
been made to create an office in the U.S. Treasury Department to
provide technical assistance for the public sector for PPPs. The U.S.
DOT already offers tool kits with expert advice. This is especially
important for pension funds that might not have the capacity for due
diligence and monitoring of investments. This way, decisions can be
made locally, but the federal government would help with capabilities (in addition to any federal grants or loans, such as what PortMiami received in TIFIA funds).

PPPs are a vessel into which money is poured from multiple
sources and which takes shape because of public priorities. It's easy
to see that there are national as well as regional strategic rationales
for a project like the PortMiami Tunnel, even in the absence of a
clear national strategy. It is also worth noting that this project, like
many counterparts, had international investors.

After largely ceding this domain to foreign firms, U.S. private
investors are getting involved. KKR, the prominent private equity
firm, recently closed a $4 billion infrastructure and energy fund
with the goal of taking infrastructure private and achieving returns
through operational efficiencies. Blackstone and the Carlyle Group,
two other prominent PE firms, have also recently formed infrastructure funds.[24] Highstar Capital, a New York–based infrastructure investor, has successfully closed three infrastructure funds
and is currently managing its fourth. General Electric's CEO Jeffrey Immelt says that GE views infrastructure as a $60 billion global
investment opportunity. David Walker, former U.S. comptroller general, suggests that pension funds could significantly increase their
investment in infrastructure assets and that these investments align
well with the long-term duration of their liabilities. Sovereign wealth
funds are another set of potential institutional investors. However,
politics enters the equation. All money is not created equal. Norway,
yes, come on in. Middle Eastern nations, not so much. Dubai Ports
has found itself unwelcome in the United States. Ownership of all or

even a portion of critical U.S. infrastructure by other nations could prove controversial.

Meridiam Infrastructure partner Joseph Aiello observes that domestic construction firms—such as Chicago's Walsh, Omaha's Kiewit, and Fluor of Irving, Texas—have grown significantly more sophisticated in structuring and executing complex projects in recent years, and the investment community continues to gain experience in novel infrastructure investments as well—to the point where American firms and funds are becoming competitive with seasoned non-U.S. players like Bouygues, Meridiam, and the European banks financing them. This growth in American capabilities could motivate U.S. firms to advocate for needed innovations.

A step toward encouraging U.S. investors is to show interest in them at the highest government levels. In early September 2014, the U.S. Treasury, Transportation, and Commerce Departments jointly hosted an Infrastructure Investment Summit. Top federal officials (including Secretary of the Treasury Jacob Lew, Secretary of Transportation Anthony Foxx, and Secretary of Commerce Penny Pritzker) spoke about new infrastructure financing methods and investor priorities to equally prominent private investors, who were collectively expected to invest about $50 billion in U.S. infrastructure over the next five years. This component of the Obama administration's Grow America initiative signals a growing federal commitment to exploring more efficient means of delivering critical infrastructure projects with increased private-sector participation. It also acknowledges the need to show the private sector that the public sector can assemble resources across the government.

How much more capital could be unlocked if there were, in fact, a national strategy and a systematic effort to remove political impediments? The research firm Preqin estimates there could be as much as $2.5 trillion globally in global private capital for infrastructure investment by 2030.[25] What conditions, they ask, are necessary to draw this private capital to the United States while also protecting the public interest—and how can leaders facilitate the investment process?

THE CASE FOR INFRASTRUCTURE BANKS

If only we could get away from politics. If only we could have a dedicated pool of funds led by experts who pick projects on their merits and fit with national priorities. If only we could attract private investors so that we don't need new taxes. We could improve deal structures and management processes so that everything would look more like the Port of Miami Tunnel and less like the Chicago Skyway. And then we could make some money for investors and the public too. If only . . .

That's the dream behind proposals for a national infrastructure bank. Such a bank is another thing that China and Europe have that the United States doesn't. The China Development Bank is said to have more than a trillion dollars in assets, according to an experienced international infrastructure deal maker. He says he was told that the bank has built fifty new airports and thousands of miles of high-speed rail and plans to invest over $50 billion in the proposed Asia Infrastructure Investment Bank. The European Investment Bank (EIB) has been the European Union's primary infrastructure financing mechanism since 1958 and in 2010 lent more than $100 billion to 460 large projects across the EU and elsewhere.[26] Maybe that's why the German autobahn and European high-speed trains are so impressive. EU member countries, twenty-seven of them, guarantee its bonds with their own credit, enabling the bank to borrow at low rates. The bank has made a small annual profit; in 2012, despite the EU monetary crisis, it posted $3.6 billion in earnings on assets of $671 billion.[27] Newer banks are less proven. The Brazilian National Development Bank lent $31.8 billion in 2010, though its economic viability cannot be determined until it comes time for these loans to be repaid.

Proponents note that more than $180 billion in private equity and pension fund capital is available globally for infrastructure equity investments and argue that an infrastructure bank could help attract a hefty portion of this capital to the United States.[28]

The infrastructure bank concept is largely untried and unproven in this country. Some experts have observed that the current TIFIA loan program, which the Miami port tunnel tapped, bears some resemblance to an infrastructure bank.[29] But Denver's Union Station officials talked about difficulties with TIFIA to Harvard Advanced Leadership Fellows in June 2014.

Removing political calculations from investment decisions would certainly be an improvement. Congressional apportionment of funds often requires that the government "spread the goodies thinly and widely," leading to overfunding of poorly conceived projects and underfunding of projects with true national significance, as my Harvard Kennedy School colleague Jose Gomez-Ibanez notes. More than 92 percent of federal highway funds are allocated by geographic formulae, without an emphasis on achieving specific economic outcomes or a return on investment—which is what makes Bridges to Nowhere conceivable. Even more egregiously, Harbor Maintenance Trust Fund grants go disproportionately to low-volume, relatively unimportant seaports—the result of a political system in which rural areas enjoy power out of proportion with their populations.[30]

The infrastructure bank idea hasn't gone very far in the United States. State infrastructure banks are possible by law in thirty-three states, but only twenty-three have actually set them up, and only a handful (such as South Carolina and Florida) are currently active, in part because an oversight in the 2009 federal transportation bill did not extend their authorization. Since 1995, these banks have arranged a minuscule total of $8.9 billion in transportation projects, not much for dozens of banks over nearly twenty years.[31]

A recent toe in the water is Chicago's Infrastructure Trust, the world's first metropolitan infrastructure bank, announced in June 2012 by Mayor Rahm Emanuel with former President Bill Clinton in attendance, and governed by a board of business executives, public officials, and labor leaders.

Accolades followed the announcement. Damon Silvers, a national labor leader for the AFL-CIO and Chicago Infrastructure Trust advisory board member, said, "We think the Chicago Infrastructure Trust is among the most creative and promising vehicles out there making an effort to deal with constraints presented by federal fiscal gridlock and austerity." Jorge Ramirez, president of the Chicago Federation of Labor and a board member, described the trust as "another layer of protection" as a kind of market test, rather than leaving it up to city officials to decide whether a deal is good. The U.S. Conference of Mayors asked Mayor Emanuel to lead its infrastructure finance committee. Former U.S. Secretary of Transportation Rodney Slater declared, "Mayor Emanuel has taken a significant step in pulling together mayors from around the country to engage about infrastructure investment."

Trust CEO Steven Beitler envisioned that the trust could someday do a sweeping range of things—from developing wind farms in Lake Michigan to getting food to inner-city food deserts. MarySue Barrett, president of the Metropolitan Planning Commission and another trust adviser, pointed to great models the trust would use: procurement protocols from Infrastructure Ontario, or communication strategies from Stockholm.[32] As an alternative financing tool, the trust could engage in joint ventures, private equity, state, federal, and private grants, as well as public-private partnerships. Funding from private sources can come in several forms, such as direct equity investment, funding via debt (municipal bonds), or Community Reinvestment Act (CRA) funds from major banks and pension funds.

Although raves for the Chicago Infrastructure Trust make it seem capable of everything short of leaping tall buildings in a single bound, it started cautiously. In November 2013, advised by investment bank Piper Jaffray, the trust tentatively approved its first project, a $25 million energy efficiency retrofit of seventy-five public buildings, to be repaid over twenty years from utility savings at an interest rate of 3.8–4.7 percent.[33] In June 2014, the trust started negotiating with private partners for a second project, to retrofit 141 city-owned swimming pools and make them more energy efficient. The work will cost $30 million to $50 million and will save the city 20 percent on energy bills tied to the pools. Local politicians worry about a possible lack of oversight and transparency. A financial expert questions the economics and reports market skepticism. But proponents hail this as the kind of innovation that can get America moving. It's a demonstration, a proof of concept, rather than a full-blown system.

Proponents hope that regional demonstrations will propel national action, even if a national infrastructure bank ends up looking like the Federal Reserve: a regional network with central governance—an idea I favor. Regions span states and are big enough to undertake significant projects while managing competition (everyone wanting an airport, for example) and being grounded enough to reflect local interests and win local support.

Proposals abound, and like PPPs, the idea of a national bank has grown to encompass many different (and competing) visions of investment; one analyst called it "a Rorschach test onto which one's preferred vision for infrastructure development is neatly grafted."[34] There is little consensus on how it would be capitalized, what specific services it would provide, how it would operate, and what institutions would govern and oversee it. In 2011, a Senate proposal, the BUILD Act, was dead in the water. As U.S. Representative Tom Rice (R-SC) noted during the America on the Move Summit, in a world where Congress has been unable to agree on a budget in years, marshaling support for a national transportation plan—and supporting mechanisms like an infrastructure bank—is easier said than done.

The other challenge is figuring out how to do all this without raising taxes—which might not be possible. Back to the private sector for sources of capital. One source, in tantalizing sight, is the foreign earnings sheltered abroad by U.S. companies. A proposed Partnership to Build America Act, championed by freshman Congressman John Delaney (D-MD), would allow U.S. corporations to repatriate foreign earnings tax-free in return for buying fifty-year "Build America" bonds, not guaranteed by the U.S. government and paying 1 percent interest, that would raise $50 billion in seed money for an "American Infrastructure Fund." Assuming a 15:1 leverage ratio, this fund could potentially provide $750 billion in financing.[35] The bill has attracted fifty cosponsors (twenty-five from each party) and the support of numerous prominent organizations, although skeptics note that any new source of government revenue will have many claimants. Perhaps there could be a bipartisan solution. Congressman Rice has expressed enthusiasm for a similar plan from House Transportation & Infrastructure Committee chairman Dave Camp, his fellow Republican, to allow repatriation of foreign earnings to fund transportation investments.

But it would not be enough to have a bank without a national vision and streamlined processes. Those billions of dollars in international private equity and pension fund capital that are supposedly available won't come to the United States unless there are changes.

TOWARD A NATIONAL VISION

Infrastructure projects in America can be so complex that they seem like a jobs act for lawyers: multiple layers of government, numerous regulatory hurdles, environmental permitting, and high levels of political uncertainty. But even lawyers understand that legal costs can grow too large relative to the size of a given project. That, plus political risks, drives away investors.

Waide Warner, who led the Project Finance Group for the international law firm Davis Polk & Wardwell from New York, looked outside the United States to ply his trade, finding it a more fruitful source of major infrastructure deals. He advised on toll road projects in Peru, power projects in Indonesia and Venezuela, and undersea fiber-optic cable projects across the Atlantic and Pacific, which required permits only to connect with local telecoms at landing points. He drew lessons later, while a Harvard Advanced Leadership Fellow in my program. Successful international infrastructure projects, he says, typically arise from strong national development policies; strategic priorities often lead to relatively streamlined approval processes and liberalized investment regimes. In the United States, in contrast, he points to "a predominance of politically motivated projects, rather than projects that reflect a coherent, long-term infrastructure plan aligned with the long-term needs of institutional investors."

In short, the American dilemma is politics, not money. We're back to the will rather than the wallet. To get more Tunnels to Tomorrow, and avoid Bridges to Nowhere, will take courageous leaders willing to show the public why infrastructure investments matter to us and why it's thus in our interest to pay a little more.

U.S. Senator Barbara Mikulski (D-MD), chair of the Senate Appropriations Committee through 2014, recalls a time earlier than my dad's municipal bond investments: the sale of War Bonds during World War II. To support the war effort, she says, "you could buy stamps and at the end you got a bond, costing maybe $18.75 and cashing to $25. I felt that I personally was helping win World War II. What would happen if we all thought about rebuilding America—that it is part of a national effort, people see things in their own community, that they are doing it. I think we can get to this . . ."

I immediately thought of Harriet in Miami; perhaps in addition to getting naming rights to the tunnel-boring machine, local Girl Scouts could have sold cookies to finance their small piece of the tunnel and the American dream. I am only half joking. Engagement creates a sense of common purpose. And those Girl Scouts could grow up to be engineers.

Mikulski also talks about how national leaders can minimize the political risk of voting for appropriations: "What if a mother of four with a minivan that costs $100 at every refill says to me, 'Barbara, you voted to raise the gas tax. I can barely afford it now.' She's going to be mad at me, unless she feels that whatever we're doing enables her to get to work, get to school, or even maybe have a mass transit option. And if you feel that a local project is giving mom or dad a job, then you'd support it. You shouldn't have to fear risking your seat if you want to rebuild America's infrastructure. Transportation is a bread-and-butter word."

That's the beauty of strategic projects, like the Miami tunnel, which create immediate jobs, solve long-standing problems, open new opportunities, and share the risk with private investors. Wallets are out there, and they can be opened. Regional, state, and local leaders are ready. But the federal government must do its part, providing a national framework and national funds. Citizens should demand it. It's all a matter of leadership.

HOW TO MOVE

Infrastructure that works for the twenty-first century could usher in a new transportation era for the nation and a renewed sense of national purpose, contributing to a bright future of shared prosperity. At least, that's the hope. Consider all there is to be gained if we lift our sights beyond fixing potholes, decrepit bridges, and broken rail track joints—all of which should be done—to reimagining how we move and the platforms we move on. We could empower pilots in the air, empower passengers on the ground, and use technology to gain choices. We could enjoy people-centered cities and could have infrastructure for mobility that is technology enabled, safe and efficient, environmentally sustainable, and opportunity focused.

For nearly six decades, America has neglected or underinvested in some critical aspects of transportation and infrastructure; fallen behind in international comparisons; generated some of the world's best technology but lagged in applying it; ceded leadership in manufacturing for some transportation sectors; and let cities deteriorate and remain divided into rich or poor, often by race. The daily delays and longer-term delays in modernization jeopardize productivity and quality of life. America needs to move. Not simply for repair and renewal of aging systems but also for reinvention of transportation and infrastructure through exciting technology in the hands of every person, to help us all become more mobile, more easily and sustainably.

The result will be an America that works better for communities and businesses, creating millions of immediate jobs and opening numerous new business opportunities. It could enhance our stature in the world as we regain the lead and become the place everyone must visit to see how the future works. Even better yet is the human dividend in lives enriched. Infrastructure could be the Big Apple demonstration that America can solve big problems again; if we can do it here, we can do it anywhere.

Sixty years ago, national defense goals shaped an infrastructure and transportation era whose legacy still defines American society. Today, we need a new national goal and rallying cry as compelling as the highways, the personal vehicles, and a race to the moon. The twenty-first-century version could be a race for mobility. Its goal: to ensure that we are the most advanced nation in getting where we need to go safely, efficiently and cost-effectively, conveniently with access for all citizens, sustainably with reduced carbon and pollution, and best able to open new opportunities for jobs and development. And while we're at it, we could reduce disparities by providing more people more access to those jobs and opportunities.

To get America moving toward building the national future we want with the infrastructure we need, we can focus on six arenas for action.

1. A NEW STORY: THE MOBILITY RACE

We need a new national narrative about where we are and where we can go. As I've found in working on change in other realms, sometimes the best way to create the future is to rewrite the story of the past. Infrastructure is something without ideology or party. Americans might disagree about how to pay for it or whether it should be handled nationally or locally, but for the most part it has no ideology or party. But it does have a history. It's important to show the ways in which that history opened big opportunities for the economy and quality of life in the past and how that foundation can be built on.

This should be a story about mobility—how physical mobility shapes social mobility, how mobility is opportunity, how the haves benefit from access to mobility, how the have-nots are constrained by lack of it. We're not just fixing infrastructure; we're building communities and a nation. It's all about mobility.

In any narrative, names matter. We should stop debating how to save the Highway Trust Fund. We should immediately rebrand it the Mobility Trust Fund and get new sources of funding for a renewed national initiative. The current trust already contains a small proportion of funding for public transit, so this, and any alternative federal funds, should broaden the name beyond the emphasis on highways. This name change would also make it possible to consider lines of funding for other forms of mobility and to find the synergies or links among them. And it would add an emphasis on technology. It might encourage links between the federal Department of Transportation and other departments. It could help the FAA and FCC— airlines and communications—develop joint strategies, given the vitally important role that information and communications technology will play in the future of infrastructure.

Every major issue facing American has a transportation infrastructure angle. A new narrative must always refer to these influences and effects. Every group and every policy should draw from the same narrative, even with variants for their own issue. The mobility sector affects safety and health—how long we live, how much we're injured, how the air we breathe affects our health, how quickly first responders reach us in an emergency, how easily we can get to health care. It influences education—length of and affordability of travel for a good education, the learning consequences of wear and tear to get to school every day. For jobs and employment, the story is huge, in direct employment, indirect employment, and economic spillovers from opportunities opened. Commercial impacts are numerous. Parking affects neighborhood businesses; delays or improvements at ports affects big businesses; employee commutes affect just about every business; and airlines and airports affect international trade opportunities, a big source of growth.

The new narrative should encompass the experiences of a single mother in Chicago who needs convenient public transit and the deal-making CEO of General Electric who needs better smartphone coverage. We should stop talking about the issues in silos, industry by industry, one transportation mode at a time, public agencies separated from private enterprise. It's all of a piece, all a connected (or less than optimally connected) system. Talking about it all together, in one story, makes it easier to remember that every national and regional conversation should include the complete range of actors and stakeholders. Not just a few people talking to others just like them, but the whole spectrum. That goes for the federal, state, and local government too. All have departments of transportation, but the DOTs should be better connected. There should be liaisons of transportation to other agencies and a call for explicit collaborations—technology and innovation with road or airport planners, and more.

Education is vitally important to reach the potential for transportation, and that should be a central part of the story. It's blindingly obvious that the workers of the future won't be toll collectors, taxi dispatchers, or parking lot space checkers in golf carts. The jobs created by intelligent transportation systems will involve data. The good news is that America recently ranked third in the UN International Telecommunication Union skills subindex. But that's not good enough. There is still a shortage of data science skills in the United States to do the work that smart-everything demands. The massive volumes of Big Data require a sophisticated set of data manipulation and IT skills. This seems like an education problem, not a transportation problem, and it is. But the opportunities in smart roads, vehicles, and networks will be constrained by the educational shortfall—another argument for an integrated national vision. Baby steps in this direction must become giant leaps. The U.S. DOT's Garrett A. Morgan Technology and Transportation Education Program, founded by former Secretary Rodney Slater to honor the African

American traffic light inventor and authorized by Congress in 2005, issues grants to improve science, technology, engineering, and math (STEM) education, focusing on transportation, with an emphasis on programs that serve women and minorities. Cardozo High School in Washington, D.C., operates STEM-focused TransTech Academy, which offers technical skills training for bus, train, and escalator maintenance and a pipeline to jobs with the Washington Metropolitan Area Transit Authority. With Federal Highway Administration support, TransTech offers a pre-engineering curriculum with college credit to equip students to go to college in transportation-related fields.[1] P-TECH-style six-year high schools in New York City, Chicago, and beyond, address Big Data skills and open doors to employment. Addressing the skills gap and the jobs problem are essential.

To educate and inform the public, the Federal DOT is already providing abundant information and indicators, as are state and local authorities. Like the census, these data tell the public what's going on in their communities and regions, and how this compares with others. There's more to be done. With the new narrative in mind and smartphones in hand, this should be enlarged to provide more national, state, and local indicators driven by human concerns, easily findable in one place, since we never know what connections people will need to make. Usable information might include average commutes or distance and time to health providers by various modes, or the vulnerability of local roads and bridges, or the best deals on Uber or the best routes for Zipcar. For commercial uses, information could encompass the efficiency of ports and airports in moving goods. When looking up flight arrival times and delays, we should also be able to get information about ground transportation departures and delays. This is just a starting point. Perhaps cities or civic associations could sponsor community hackathons in every region, in which a range of organizations contribute data they wish to have and share.

The new narrative should not abandon what we have—we'll still have cars, suburbia is still growing—but it should reframe the characters to rebalance the story and become a story about moving into the future. Priorities should be clear. Rail and mass transit should play a leading role. Streets should be reinvented to make them people friendly. We should flip the emphasis in policies from suburbia to cities, from individual action to technology connections, from single modes of transportation to intermodal connections among them. And while we're at it, we should revisit every policy created more than twenty-five years ago in order to modernize it and bring it up to date, such as international Open Skies agreements.

Infrastructure for mobility should be the new national defense strategy as well as our race for the future. If we're stuck in traffic, we can't be leaders in the world. Or if defense metaphors are not to the public's taste, the story can use health metaphors. We must ensure that the circulatory system of society has unclogged arteries so that our national lifeblood can flow.

2. A NATIONAL FRAMEWORK IMPLEMENTED REGIONALLY

A twenty-first-century vision is not a singular strategy or a centrally administered plan. The vision should identify principles—the quintuple wins are a good starting point—but then facilitate action at the local or regional level. We need a national framework and federal funding to ensure consistency of standards—auto safety, aviation regulations—and also to ensure that investments are smart and strategic, well thought through and in the national interest—not pork, not proposed bridges to nowhere or underutilized urban people movers, but investments in the future that produce new value for the public.

A strategic national framework could jump-start action on transportation infrastructure by reducing uncertainty and making the priorities and rules clear, thereby opening the way for innovation. Long-term investors, whether pension funds, sovereign wealth funds, private equity, or local taxpayers, are waiting for a framework they can count on for the longer term.

Although I'm urging us to see many things as connected, that doesn't mean that we can work on everything at once. As in any effort at system change, we have to get started with focused steps, a few projects at a time. But those steps will turn into isolated, disconnected, low-payoff fragments unless the big vision and organizing principles are kept clearly in mind at all times, so that the pieces add up, get multipliers, and have impact. Multipliers include the spillover benefits when bus rapid transit also gets water mains repaired, or renewed bridges have sensors added, or untangling railroad grade crossings is accompanied by parks and bicycle paths, or light rail systems add stations that spur neighborhood development, or a port tunnel maximizes future jobs as well as downtown residences and amenities for pedestrians.

Within a national strategic framework, implementation should be led by regional, often metropolitan, planning groups, with the flexibility to direct funding. Local officials receiving federal funds should similarly have flexibility to respond to local conditions, as long as they remain true to the organizing principles. Federal funds can come in the form of competitions for regional projects that fit strategic priorities, challenge grants that must be matched by other financial sources, or seed grants for early stages that must demonstrate the potential for results in order to move on to the next phase of support. This helps national priorities to be reflected in regional actions without overly constraining regional choices.

What is a region? Let it be self-defined and issue flexible. Between central cities, suburbs, states, broader territories, and all the communities within them, there is ample scope for varying definitions of regions depending on the issues. Regional boundaries should be self-defined because they should be fluid, not rigid jurisdictions; they should be organized on the basis of the particular issue, perhaps crossing state lines, or perhaps encompassing mainly a city.

Regions are important, as the metropolitan planning advocates know. In fact, regional bright spots have appeared throughout this book. Some projects have already gone from concept to operation, while others must be done in phases because of their sheer magnitude. Some are driven by entrepreneurs starting services, apps, or data analytics businesses, while others are big public works projects. What they have in common is the obvious fact that they are all associated with specific places, either because they're addressing that place or because their implementation depends on local acceptance. All politics is local, said late Speaker of the House Tip O'Neill in a widely quoted phrase. Today, with transportation in mind, we might say "regional" instead. Same difference. Both imply that infrastructure projects must get buy-in one neighborhood at a time, as Beverly Scott learned from her long career at the helm of public transit authorities.

State, local, and regional leadership drives progress, even when federal programs set priorities and provide funding. And the particulars of what progress means and what projects achieve it are based on local conditions, needs, and preferences, which vary greatly from place to place. Local conditions include a diversity of communities, which may be divided by race and ethnicity, social class, dominant employers, or functions within a region.

Thinking locally and regionally doesn't mean retreating into isolated territories. As part of regional plans, we should encourage corridor strategies that define intercity and cross-regional pathways—the territories for mobility and trade. The Boston–Washington corridor in the Northeast has been reinforced as a pathway through frequent rail and air connections dating back a century or more. Now the State of California has ambitious plans for high-speed rail; private-sector entrepreneurs are planning high-speed rail connections in Texas and Florida. Perhaps the next Transcontinental Railroad could be an enhanced North–South corridor through the Midwest, connecting more routes from Canada to Mexico, growing a lucrative North American free-trade zone. States are rallying to work across boundaries. The West Coast Infrastructure Exchange, an Oregon-led partnership between California, Oregon, Washington, and British

Columbia in Canada, formed in late 2012, seeks innovative methods to finance and facilitate infrastructure development for the region's competitiveness, including identifying critical regional projects. On the other coast, Governor Martin O'Malley of Maryland, with the support of the Clinton Global Initiative, announced in June 2014 a new Mid-Atlantic Infrastructure Exchange to expand cross-state collaboration.[2]

Since regions are fluid and cross boundaries of political jurisdictions, they need a different kind of leadership and governance. Mayors, governors, and county officials must clearly be involved and cooperate—Mayor Rahm Emanuel of Chicago can't unilaterally direct what goes on at O'Hare Airport, because it sits in two counties. Public authorities for ports/airports and transit span jurisdictions—the Port Authority of New York and New Jersey, which also runs some bridges—with their charters indicating how many board seats are appointed by which public official. Added to this are private-sector groups and associations that take on a civic mission and look across jurisdictions to determine what's best for the region—or themselves—and working with—or lobbying—public officials.

Numerous examples throughout this book make clear the important role of regional coalitions at every point in the development of new or improved transportation and infrastructure. Denver's Union Station had the backing of a wide set of stakeholders. Atlanta's Hartsfield Airport modernization program has featured numerous cross-sector groups drawn from many organizations. The first phase of modernization of Chicago's O'Hare Airport was led by a business group pushing a reluctant Mayor Daley.

Business leadership is often a critical component of regional coalitions, but so are large nonprofit organizations such as universities and hospital systems. Unlike footloose companies that can move anywhere, educational, cultural, and health care organizations are more likely to be rooted in a community and have significant stakes in the state of transportation and infrastructure. Private-sector groups are thus important not only to the buy-in but also to the vision and ongoing operations. Public-private multistakeholder coalitions can set priorities, find synergies across efforts, and leverage existing assets. These are also entities closer to the public than the distant federal government is; when people run into one another frequently, they tilt toward being more responsive to local needs and less partisan.

Of course, we must make sure that such coalitions don't become cabals hoarding power; criteria for any funding that involves them should include diversity and representation across industries and community groups. And we must educate managers and leaders about how to create coalitions and work cooperatively as members. The best coalitions are led by concerned champions with a long-term view who can find common ground in improvements in the public interest—public goods rather than benefits only to some. The collaboration of otherwise competing Class I railroads with one another, and then with civic groups, to envision Chicago's CREATE projects is a positive example. The numerous groups lined up behind Chicago's Bus Rapid Transit, and the widespread participation of residents in open public forums built support for that big change. A chance to contribute builds commitment. An inspiring shared vision helps build the fortitude to persist through the inevitable ups and downs, what I call the messy middles of change.

At a time when national inaction contributes to high uncertainty and high anxiety, regional leaders are a good bet for generating strategic thinking. If regional coalitions can push their own elected officials for action, they can potentially have an impact on national officials, including on Capitol Hill. Oregon's experiments with vehicle-miles-traveled fees for highways, instead of gas taxes, using volunteers in a pilot test, has led to a bill now in committee in Congress. It might be stalled for now, but it's on the agenda, with a robust regional demonstration to give it credibility.

In short, it takes more than a village to create change. It takes a cross-sector, multistakeholder coalition. We must continue to get better at forming and managing these coalitions. American strength

at innovation must be matched by becoming equally strong at collaboration, which currently needs improvement. In a nation that
can't force action from the center and doesn't believe in it anyway,
regional coalitions are essential to our democracy and one of our
best hopes for getting things done.

3. PUBLIC-PRIVATE PARTNERSHIPS

There's a tired debate in America about the private sector versus
the public sector, and who does things better. However, a tour of the
history, funding, ownership, and governance of various transportation modes shows that they have been neither fully public entities nor fully private businesses driven by profit. We see oscillation
between private and public ownership and governance throughout
the history of railroads. Freight rail was nationalized several times
and then privatized, while national passenger rail, which uses many
of the same tracks, is still a government entity, but expected to operate like a private business anyway. The first subway was privately
owned and operated for fifty years, as were many municipal bus systems, before it was bought out and taken over by the public sector,
by way of a jurisdiction-spanning, quasi-independent public transit
authority. The recent federal bailouts of General Motors and Chrysler were temporary nationalizations of major corporations.

It's also striking that we don't expect ownership or governance to
shift from business to government in the United States, because we
lack the vocabulary to describe this. We have a term for selling public assets to the private sector: "privatization." What's the equivalent
term for state and local governments' acquiring formerly privately
owned entities? "Publicization" doesn't do it. Taking something public or "going public" means a stock market offering; it doesn't mean
turning it over to the public sector. Yet, that sometimes happens.

The fluctuation between public and private can also occur in
a short time frame. In the case of bike sharing, New York City was

determined not to put in any money, seeking a private-sector sponsor and giving the contract for operating it to a private-sector company. Chicago began with public funds and used the same private company for operations. And there's sometimes confusion about what a public asset is. Does a local government that owns a street also own information about whether a parking space on that street is available? Can a city forbid private buses from stopping at public bus stops? These are live controversies.

We've also seen examples of problems with outright privatization, or sale of public assets with no public involvement thereafter. Privatization has historically worked in some industries when an operating entity could be broken off, such as the conversion of municipal electric power plants into for-profit businesses. But it's hard to sell airports; the FAA has had a special program for this for many years, with little to show for it. Toll roads and parking meters have also had their privatization problems. Mayor Emanuel's administration in Chicago pressed to renegotiate an unfavorable parking meter deal constructed by the previous administration, which appeared to be seeking immediate cash rather than considering the longer-term consequences.

Public-private partnerships can get the best of both worlds. PPPs, such as the Port of Miami Tunnel project, completed and operated by MAT Concessionaire, is a clear model. This way of getting more and better infrastructure projects should be encouraged. A low-cost national educational and technical assistance program for potential investors, who are sometimes daunted by the red tape, could stimulate private-sector investor interest and capability for public-private partnerships. Training and education for local governments at the regional level should be available to help build public-sector capabilities for collaboration and coordination across sectors. The White House could mount a series of convenings across the country. DOTs, in collaboration with experienced private sector financial groups, could develop tool kits and technical assistance both nationally and regionally.

We must also encourage more U.S. companies to invest in and get involved with infrastructure projects, rather than leaving them to the public sector or abdicating a role to foreign investors. If American businesses have a bigger stake in the issues, they can be more knowledgeable participants in planning coalitions. There's another virtue of emphasizing regional coalitions and supporting them. Public officials can come and go with elections and a succession of administrations, creating uncertainty about whether projects will survive or be changed; in Puerto Rico a new governor had to pledge to continue a privatization of the San Juan airport started under his predecessor or lose the funding, even though he was initially opposed to it. In a global economy, corporations can move operations and headquarters to gain financial advantages (with "inversion" the latest controversial practice), but it is also the case that business leaders, or the companies they represent, can look at a longer term than the election cycle.

Despite the difficulty of getting traction for a national infrastructure bank, a Federal Reserve–like system of regional infrastructure banks that offer seed funding to encourage private investments in strategic projects could be viable. We should learn from state infrastructure banks and local initiatives and keep the idea in discussion nationally. Just because legislative proposals have been shot down in Congress several times in the last few years shouldn't mean that proponents give up. After all, Jose Abreu in South Florida waited thirty years for his dream of a Miami port tunnel to become reality.

4. ENTREPRENEURS AND TECHNOLOGY TALENT AT THE TABLE

Technology is not a cure-all, and it can hit obstacles that call viability into question, as a NextGen program has. But it is certainly a grand enabler. Clearly, there are many unresolved issues about the emergence of new technologies, with rules still to be written. But the potential of technology, like roadbed sensors or NextGen air traffic control, is enormous for the quintuple wins. The imagination of innovators who build applications off telecom platforms such as 4G LTE can reshape mobility as they envision smarter infrastructure; that's why the giant Verizon has opened innovation labs to innovators in big companies and start-ups alike. The Volpe Center's partnership for tests of automated cars brings together the research arm of the U.S. Department of Transportation with a major academic institution, the University of Michigan, and just about the entire auto industry. This is another example of stimulating innovation across sectors by joining forces. For potential game changers such as self-driving vehicles, entrepreneurs and regulators must get into alignment.

Connecting visionaries across silos is especially important because innovations don't generally fall into the categories we've established in our minds and classification schemes. That's what makes them innovations. I've discussed cars and airplanes in different chapters of this book, but fusion products are already under development. Terrafugia, started in 2006 and initially funded by founder Carl Dietrich's $30,000 prize in an MIT student competition, has developed viable flying cars. My team saw them, although we didn't try them yet. Terrafugia has advance orders for more than a hundred of them at $300,000 each. It's still classified as an airplane, but it is just as potentially feasible for road use, and some observers call Terrafugia the only registered auto manufacturer in Massachusetts. So what is it, car or plane? It's an example of thinking not just outside the box, in a common metaphor for creativity, but outside the building—well beyond current institutions and sectors.

Tapping outside-the-building thinking suggests that those creating technologies of the future should be at the table discussing infrastructure and transportation agendas with leaders in many transportation industries. Techies, some of them antiestablishment millennials, should be included in every convening about transportation, whether by the White House or by regional civic associations. They should be a part of every federal or state grant program for tests of new possibilities. Perhaps someday there will be flying cars taking off from highways or Hyperloops whizzing through tunnels, so it could be a good idea to invite Terrafugia founder Dietrich or Elon Musk of Tesla, SpaceX, and the hyperloop. Had Segway founder Dean Kamen been at the table with government officials, the Segway story might have come out differently.

But we don't have to look for large blockbusters or transformative breakthroughs to see the importance of including these voices in transportation and infrastructure strategies. Incremental innovations can be springboards for progress, as is evident in some rail projects, such as the Virginia Avenue Tunnel and even Amtrak's ability to get slightly higher speed in Pennsylvania's Keystone Corridor at relatively low cost. There are also entrepreneurs within established companies who push boundaries and imagine new ways to enhance mobility. Creative airline pilots who improvise have knowledge that should be shared more widely than just with the FAA, although the FAA is certainly the point of first approval. Data scientists and meteorologists from the Weather Company are developing data analytics that could be applied more widely than just to air transportation; connected cars, railroads, and public transit could benefit.

The entrepreneurial mode of getting things done is to just do it. Whether Zipcar, Uber, Bridj, or parking apps, start-up companies and their founders create new models quickly, sometimes so focused on the venture that they don't think about the consequences, or brush off the rules as not important. But because they might also bump up against the public sector and regulations at nearly every turn, including them in public forums also means that they learn about the concerns, such as safety or use of public bus stops, and build relationships across the sectors. This would help Bridj get support for picking people up in central locations in its home city, or Uber's founder avoid the jail time he jokes about.

In short, to move America forward, it's important to engage entrepreneurs and emerging leaders in the quest for mobility. We should publicize the needs and business opportunities. With public encouragement, civic groups or technology proponents should run innovation contests. Together with the media, they can make transportation and infrastructure cool on college campuses, well beyond the relatively small number of civil engineering departments, reaching liberal arts students interested in cities, business school students interested in commercial possibilities, and computer scientists building data platforms. In this realm, too, we should use seed funds and challenge grants to stimulate and find promising ideas; the relatively modest funds could come from federal or state DOTs or be matched by venture capitalists. Contests can help, too, such as the U.S. DOT Data Innovation Challenge prize won by RideScout's Joseph Kopser. Other existing federal innovation tools, such as Small Business Innovation Research grants, could explicitly include transportation and infrastructure. But whatever else, we must listen to technology voices, including those of young techies, in every metropolitan planning council, civic association, and government advisory council. Their views and involvement are critical to building the future.

5. COURAGEOUS LEADERS
WITH A COMPELLING VISION

Thank goodness for American entrepreneurship and technological strength to get us started. But to scale innovations and reap the benefits from them means changing institutions, shifting paradigms, and aligning diverse interests. That requires leadership. And of course, as we hear endlessly, it should be bold and courageous. If not in this area of national importance with so little ideology involved, then where?

Those holding public office think they get elected by pandering to short-term interests and never mentioning raising taxes, and they're probably partly right. Here's what one very high-level public office-holder told me off the record: "No one who wants to be popular and elected will stand in front of voters and say, 'We're running out of living off the last generation's investment. You've had a free ride, now it's time to pay up. You've been enjoying your good life at the expense of your own children and grandchildren.'" He thinks that's exactly what should be said. But he also agrees with those who think that there's no constituency for the future. The future doesn't vote. I'm not sure that's true, as I'll argue shortly.

This kind of thinking makes us yearn for the next Abraham Lincoln for the White House or Tip O'Neill for Congress. Recall that Lincoln never let go of the vision for the Transcontinental Railroad and did the deals that made it possible, while freeing the slaves and holding a fragile nation together during the Civil War. As Speaker of the House at the helm of Congress, Tip O'Neill made mass public transit part of highway funding. These are two examples of combining vision with great political skills to sell the vision.

But even without nostalgia for great figures of history, there are several things we can do to get closer to that kind of leadership for change.

- Vote for those with a bipartisan transportation and infrastructure agenda, who connect all three R's of repair, renewal, and reinvention. Vote for bond issues and municipal levies. It's actually possible. Voters can often be trusted to do the right thing. In Florida, Miami-Dade County voters twice authorized a property tax increment to fund children's services, even in an antitax, recession-ridden environment, and they voted for a sales tax increment to fund public transit. In the transit case, voters did the right thing, but officials diverted the funds to road repairs until community activists lobbied to reverse that. Making noise gets the attention of elected officials and can change their behavior, or encourage different leaders to emerge.

- Look to mayors and governors. National is ugly, but local is beautiful, I've said, to describe the current political scene. While Congress is gridlocked and essential federal funds are threatened by budget cuts or government shutdowns, governors and mayors can get things done. Governor Jerry Brown of California is moving his state toward the kind of high-speed rail impossible to get moving in the rest of the country. And although mayors might be cash-strapped, they are crafting strategies that help attract other sources of capital, whether from foundations, civic associations, or private investors, as well as competing for state and federal funds by developing important strategic projects, such as getting air pollution mitigation grants for bicycles. They are producing small wins, but that's a start. Mayors tend to operate on a non-partisan basis and are often elected that way. They experience the truth that infrastructure has no ideology. Mayors are judged on results such as whether first responders can get through clogged streets fast enough to deal with crime or fight fires.

- Include the private sector. Business leaders know that infrastructure is essential to their businesses. They even favor more and better public transit, if we can believe the views they expressed on the recent Harvard Business School U.S. competitiveness survey. Some companies take an active civic role, and their leaders can be enlisted, if they are not already heading civic coalitions pushing for infrastructure. They can give political cover to elected officials, to increase their courage.

Institutional change arouses opposition from those benefiting from the status quo. We've seen this even with respect to parking Zipcars, adding bicycle-sharing stations, or creating ride sharing that competes with taxis. That is one reason for smart strategies and clarity about purpose. To overcome resistance to change, leaders must inspire by invoking a higher purpose, a vision that people can rally behind. And leaders must themselves believe that it's possible to do something positive that will make a difference for the nation and for people's lives. Otherwise we would have to issue a new children's book called *The Little Engine That Can't and Never Will*. That is not the American spirit. We must remember that we're the have-done and can-do nation.

6. INFORMED AND EMPOWERED CITIZENS

Infrastructure is one of those complex problems that can seem overwhelming and very technical. Its by-products can also make us feel out of control and likely to turn away from the topic, making it easier for do-nothing folks to do nothing. Traffic congestion breeds anger—Americans are famous for road rage, with the reptilian parts of the brain in high gear—and there's even more anger if told that the situation will take years to fix. When faced with things that seem to be out of our control, we tend to give up, go passive, say that nothing can be done, and make it someone else's problem. It's all their fault. That increases anger and helplessness, and it doesn't do anything to improve the situation.

Information is power, and arming people with information is a good way to engage the more rational parts of their brains and get them engaged. Information calms people down long enough to focus on a bigger picture. Recall the story of how an American Airlines pilot dealt with a planeload of upset and potentially rebellious passengers facing delays: he gave them information. More than that, information can empower people to put daily frustrations into a larger context, which helps them support an action agenda. And information can give people things to do. Passivity depresses, but activity energizes. That's why I often say that any action can be better than none, as long as it has a plausible basis in knowledge.

It's impressive to see the empowering effects of new information and communications technologies. They enable machine-to-machine communications that can help with adjustments or someday soon save vehicles from crashing without human intervention. We humans are not counted out of the equation; in fact, our roles can be bigger and more important. Apps for pothole finding and reporting are in use in several cities, involving residents to get and convey information. The Weather Company's Weather Underground network of 25,000 personal weather stations in the United States and 10,000 nearby augment National Weather Service sources to permit ever-better forecasts for ever-better planning, for everyday life as well as airline pilots and air traffic controllers, who can use new real-time communication systems to make joint decisions about the data. And by the way, those personal weather stations are operated by volunteers who love weather. The U.S. Commerce Department, which includes the National Weather Service, is considering further opening more of its own data for private use, because it recognizes the empowering effects of data.

Citizens can be empowered and inspired to create change by involving and informing them. In Oregon, Earl Blumenauer, a Portland city council member and commissioner of public works and now a long-serving congressman, worked with others to create a free course about transportation issues for citizens, to extend their understanding and also enable them to teach their neighbors. A dozen years later, classes are still filling. Progress, he says, comes down to "making the issues seem real to people, and giving them a sense that there is something they can do, building the coalitions and getting them to stop battling each other." This local program could be replicated elsewhere.

Empowered citizens are members of communities and often involved with nonprofit community-based organizations. We should ask them how transportation and infrastructure affects those they serve and how it can be improved, and join forces to advocate for that. Informed citizens are also parents and grandparents who can help educate children, beyond the basics of car seats, seat belts, or safety on bicycles and on public transit. We should talk to children about the toy trucks, cars, miniature railroads, and planes they like and mention the infrastructure each requires, the consequences of too many of them, and the alternatives. Safety is an especially big deal for children. The wheels on the bus go round and round, children sing—until the brakes fail and hurt someone because the buses are too old. Transportation and infrastructure should be a core parenting and family conversation—using simpler words, of course.

In addition to voting, empowered citizens can vote with their feet. They can support bicycle-recycling centers that make bikes affordable in poor neighborhoods and teach biking skills. If a city extends subway service extra hours as a test of longer service, then ride it, to help prove there's a market. Enlist the local media, which give a great deal of coverage to transportation and infrastructure as it is, but often only in the negative: traffic fatalities, flight delays, a water-filled hole in the street swallowing an SUV. Stress the positive possibilities.

America's infrastructure problems are too important to be left to professional planners alone, however talented they are. Many small steps informed by a clear agenda and strategic priorities can add up to propel larger change. Taking a step, even a small step, and seeing a short-term success builds confidence that change is possible. Like walking, doing something every day is good for mental health.

MOVE

It's time to think differently. It's time to shape new expectations about who we are and what we can become. Let's take the idea that mobility is opportunity, and that we're in a mobility race, as an impetus to find common purpose. Let's start a national conversation with dialogues in every region to build support for action.

And let's not delay. We don't want to be late for our appointment with the future.

NOTES

A word about sources and citations: quotations and attributions to individuals without specific citations come from firsthand interviews, field visits, or comments made on the record at the America on the Move Summit—and thus are original to this research and this book. Only published sources are cited in the notes. Information that is widely known from many sources, such as widely reported speeches by leaders, multiple press accounts of the same phenomenon, or known to me from other research and observations, is used without specific references. Notes are consolidated within paragraphs. A single note may contain several references, covering sources for multiple facts throughout the paragraph.

1. Stuck on the Way to the Future

1 David Schrank, Bill Eisele, and Tim Lomax, *Urban Mobility Report 2012*, Texas A&M Transportation Institute, Texas A&M University, Dec. 2012, p. 1, http://d2dt15nnlpfr0r.cloudfront.net/tti.tamu.edu/documents/mobility-report-2012.pdf, accessed Aug. 2014.

2 Jonathan Levy, Jonathan Buonocore, and Katherine von Stackelberg, "Evaluation of the Public Health Impacts of Traffic Congestion: A Health Risk Assessment," *Environmental Health* 9 (2010), doi:10.1186/1476–069X-9–65.

3 National Transportation Safety Board, "Preliminary Report, Accident Number DCA13MR003," June 4, 2013, http://www.ntsb.gov/investigations/2013/bridgeport_ct/Bridgeport_CT_10_day_Preliminary_Report06042013.pdf, accessed Dec. 2013.

4 Associated Press, "Thousands of U.S. Bridges Vulnerable to Collapse," CBS News, May 25, 2013, http://www.cbsnews.com/8301-201_162-57586213/, accessed Nov. 2013.

5 Federal Highway Administration, "Deficient Bridges by State and Highway System," March 2014, http://www.fhwa.dot.gov/bridge/deficient.cfm, accessed Aug. 2014.

6 Stephen Lee Davis and David Goldberg, *The Fix We're In For: The State of Our Nation's Bridges 2013*, Transportation for America, June 2013, p. 2, http://t4america.org/docs/bridgereport2013/2013BridgeReport.pdf, accessed Aug. 2014.

7 Michael Ball et al., *Total Impact Delay Study: A Comprehensive Assessment of the Costs and Impacts of Flight Delay,*" Final Report, National Center of Excellence for Aviation Operations Research (NEXTOR), Oct. 2010, p. 14, http://www.isr.umd.edu/NEXTOR/pubs/TDI_Report_Final_10_18_10_V3.pdf.

8 Lawrence Blincoe, Ted Miller, Eduard Zaloshnja, and Bruce Lawrence, *The Economic and Societal Impact of Motor Vehicle Crashes, 2010*, National Highway Traffic Safety Administration, May 2014, http://www-nrd.nhtsa.dot.gov/Pubs/812013.pdf, accessed Aug. 2014.

9 Ibid., pp. 1–4.

10 U.S. Environmental Protection Agency (EPA), Federal Highway Administration, *Environmental Justice Emerging Trends and Best Practices Guidebook*, Dec. 2011, sec.1, http://www.fhwa.dot.gov/environment/environmental_justice/resources/guidebook/guidebook01.cfm. EPA, *Inventory of U.S. Greenhouse Gas Emissions and Sinks: 1990–2012*, April 15, 2014, p. ES-11, table ES-3, http://www.epa.gov/climatechange/Downloads/ghgemissions/US-GHG-Inventory-2014-Main-Text.pdf, accessed Aug. 2014.

11 EPA, *Inventory*, p. 2–26.

12 Schrank, Eisele, and Lomax, *Urban Mobility Report 2012*, p. 13.

13 Richard Weingroff, "Federal-Aid Highway Act of 1956: Creating the Interstate System," *Public Roads* 60, no. 1 (Sept. 1996), http://www.fhwa.dot.gov/publications/publicroads/96summer/p96su10.cfm, accessed June 2014.

14 Christopher H. Wells, "Fueling the Boom: Gasoline Taxes, Invisibility, and the Growth of American Highway Infrastructure, 1919–1956," *Journal of American History* 99, no. 1 (2012): 79–80, doi:10.1093/jahist/jas001.

15 Federal Highway Administration, "Dwight D. Eisenhower National System of Interstate and Defense Highways," Jan. 2014, http://www.fhwa.dot.gov/programadmin/interstate.cfm, accessed June 2014.

16 Federal Highway Administration, Office of Planning, Environment, and Reality, *Environmental Justice Emerging Trends and Best Practices Guidebook*, Nov. 2011, p. 11, http://www.fhwa.dot.gov/environment/environmental_justice/resources/guidebook/ejguidebook110111.pdf, accessed Aug. 2014.

17 Elise Gould et al., "What Families Need to Get By," Report, Economic Policy Institute, July 3, 2014, http://www.epi.org/publication/ib368-basic-family-budgets/, accessed July 2014.

18 U.S. Department of Transportation, Bureau of Transportation Statistics, *The Changing Face of Transportation*, 2000, http://www.rita.dot.gov/bts/sites/rita.dot.gov.bts/files/publications/the_changing_face_of_transportation/pdf/entire.pdf, accessed Aug. 2014.

2. On the Rails

1 Central Japan Railway Company, "Outline: History and Overview of the Tokaido Shinkansen," 2014, http://english.jr-central.co.jp/about/outline.html, accessed June 2014. Tony Jin, "China High Speed Train Development and Investment," *China Perspective*, Dec. 27, 2012, http://www.thechinaperspective.com/articles/chinahighspeedr-9905/, accessed June 2014. Idem, "Reliability," 2014, http://english.jr-central.co.jp/about/reliability.html, accessed June 2014. Roderick A. Smith, "The Japanese Shinkansen," *Journal of Transport History* 24, no. 2 (2003): 222–36.

2 Jon Stewart, "Maglevs: The Floating Future of Trains?," BBC, May 6, 2012, http://www.bbc.com/future/story/20120504-the-floating-future-of-trains, accessed June 2014. Transrapid International, Siemens/ThyssenKrupp, "The Future Is Already Here: The Transrapid Maglev System in Shanghai," April 2010, http://www.transrapid.de/pdf/TRI_shg_10_04_E.pdf, accessed June 2014. "Linimo Hovers Close to Take-off," *Railway Gazette*, Sept. 1, 2004, http://www.railway gazette.com/news/single-view/view/linimo-hovers-close-to-take-off.html, accessed June 2014. "Power without Wheels," *Korea Herald*, May 21, 2014, http://www.koreaherald.com/view.php?ud=20140521001593, accessed June 2014.

3 Jean-Pierre Arduin and Jincheng Ni, "French TGV Network Development," *Japan Railway & Transport Review*, no. 40 (March 2005): 22–28, http://www.jrtr.net/jrtr40/pdf/f22_ard.pdf, accessed June 2014. "High Speed Lines in the World," International Union of Railroads, Sept. 2014, http://www.uic.org/IMG/pdf/20140901_high_speed_lines_in_the_world.pdf, accessed June 2014. "The World's Longest High-Speed Railway Lines," Railway Technology, Dec. 20, 2013, http://www.railway-technology.com/features/featurethe-worlds-longest-high-speed-railway-lines-4149752/, accessed June 2014.

4 James Dao, "Acela, Built to Be Rail's Savior, Bedevils Amtrak at Every Turn," *New York Times*, April 24, 2005, http://www.nytimes.com/2005/04/24/national/24acela.html, accessed June 2014.

5 Federal Railroad Administration, Office of Safety Analysis, "Highway-Rail Incidents Reported on Form FRA F 6180–57," table 5.5: Consolidated Hwy Rail Accident Incident.

6 Pullman Railway Journeys, "A-Class Accommodation," http://www.travelpullman.com/accommodations, accessed Dec. 2013.

7 "Infrastructure: Back on Track," *Economist*, April 13, 2013, http://www.economist.com/news/business/21576136-quiet-success-americas-freight-railways-back-track, accessed Dec. 2013.

8 "The Railroads," *American Experience: Andrew Carnegie*, PBS, 2009, http://www.pbs.org/wgbh/amex/carnegie/sfeature/m_run.html, accessed Aug. 2014.

9 "The Story of a Great Monopoly," *Atlantic*, March 1881, http://www.theatlantic.com/magazine/archive/1881/03/the-story-of-a-great-monopoly/306019/, accessed Aug. 2014. Alfred Chandler, *Bonds of Enterprise: John Murray Forbes and Western Development in America's Railway Age* (Cambridge: Harvard University Press, 1984), p. 177.

10 Joel Palley, "Impact of the Staggers Rail Act of 1980," Federal Railroad Administration, March 2011, http://www.fra.dot.gov/eLib/Details/L03012, accessed Dec. 2013.

11 Statista,"Rail Industry Dossier 2012," pp. 16, 17, 24, www.statista.com, accessed Dec. 2013. Federal Railroad Administration, "Freight Rail Today," 2013, http://www.fra.dot.gov/Page/P0362, accessed Dec. 2013.

12 "Uncoupling the Trains," *Economist*, Nov. 10, 2012, http://www.economist.com/news/business/21565982-plans-liberalise-europes-rail-services-run-opposition-uncoupling-trains, accessed Dec. 2013. "The Quest for a Ryanair of Rail Freight," *Economist*, Aug. 17, 2013, http://www.economist.com/news/business/21583703-eu-wants-more-goods-be-moved-train-progress-slow-quest-ryanair-rail, accessed Dec. 2013.

13 Association of American Railroads, "Class I Railroad Statistics," July 9, 2013, https://www.aar.org/StatisticsAndPublications/Documents/AAR-Stats-2013–07–09.pdf, accessed Dec. 2013. MarketLine, "MarketLine Industry Profile: Road and Rail in the United States," March 2013, p. 13. Joel Palley, "Freight Railroads Background," Report, Federal Railroad Administration, April 2013, http://www.fra.dot.gov/eLib/Details/L03011, accessed Dec. 2013.

14 Association of American Railroads, "Class I Railroad Statistics." Statista, "Rail Industry Dossier 2012," pp. 25–26.

15 Jad Mouward and Elizabeth A. Harris, "When Mileage Means Money," *New York Times*, June 17, 2014, pp. B1, B9.

16 CSX's score was based on the company's responses to the questionnaire from the Carbon Disclosure Project, an international, not-for-profit organization representing 655 institutional investors with $78 trillion in assets that provides the only global system for companies and cities to measure, disclose, manage, and share vital environmental information, which focuses on greenhouse gas emissions, emissions reduction targets, and the risks and opportunities associated with climate change. CSX Corporation Inc., "About CSX—Awards and Recognition," http://www.csx.com/index.cfm/about-csx/awards-and-recognition/, accessed June 17, 2014.

17 Catherine Byerly, "Fortune Magazine Ranks CSX among World's Most Admired Companies in Industry," *Jacksonville Business Journal*, Feb. 28, 2014, http://www.bizjournals.com/jacksonville/news/2014/02/28/fortune-magazine-ranks-csx-among.html, accessed June 17, 2014. "The DiversityInc Top 10 Companies for Veterans," *DiversityInc*, http://www.diversityinc.com/top-10-companies-veterans/, accessed Aug. 2014.

18 John Schwartz, "Chicago Train Congestion Slows Whole Country," *New York Times*, May 7, 2012, http://www.nytimes.com/2012/05/08/us/chicago-train-congestion-slows-whole-country.html.

19 Chicago Region Environmental and Transportation Efficiency Program, "Common Operational Picture Fact Sheet," Dec. 2010, http://www.createprogram.org/factsheets/Common_Operational_Picture.pdf, accessed Aug. 2014. Idem, "Final Feasibility Plan," Aug, 2005, http://www.createprogram.org/feasibility

.htm, accessed Dec. 2013. Idem, "Overall Project Status Summary," May 2013, http://www.createprogram.org/linked_files/status_map.pdf, accessed Dec. 2013. Alex Goldmark, "PICS: Largest Truss Bridge Ever Moved into Place Fully Assembled," WNYC, Aug. 27, 2012, http://www.wnyc.org/blogs/transportation-nation/2012/aug/27/largest-truss-bridge-moved-while-fully-assembled-chi cago/, accessed Dec. 2013. Richard Wronski, "Chicago Rail Program a Success, but Future Funding in Doubt, Officials Say," *Chicago Tribune*, June 11, 2013, http://articles.chicagotribune.com/2013-06-11/news/ct-met-railroad-bottle-necks-20130611_1_rail-crossings-create-program-englewood-flyover, accessed Dec. 2013.

20 Federal Railroad Administration, *Comparative Evaluation of Rail and Truck Fuel Efficiency on Competitive Corridors*, Final Report, Office of Policy and Communications, Nov. 19, 2009, http://ntl.bts.gov/lib/31000/31800/31897/Compar ative_Evaluation_Rail_Truck_Fuel_Efficiency.pdf, accessed Aug. 2014.

21 National Gateway, CSX Corporation, "Projects," 2009, http://www.national gateway.org/projects, accessed Aug. 2014.

22 "$77,000,000 Loan Granted to P.R.R.," *New York Times*, Dec. 30, 1933, http://query.nytimes.com/mem/archive/pdf?res=FB0B17F63E5516738DDDA90B 94DA415B838FF1D3, accessed Dec. 2013. National Gateway, CSX Corporation, "National Gateway Receives Federal Stimulus Funds through TIGER Program," Feb. 17, 2010, http://www.nationalgateway.org/news-resources/press-releases/2010/national-gateway-receives-federal-stimulus-funds-through-tiger-pr, accessed Aug. 2014.

23 National Transportation Safety Board, "Preliminary Report, Accident Number DCA13MR003," June 4, 2013, http://www.ntsb.gov/investigations/2013/bridgeport_ct/Bridgeport_CT_10_day_Preliminary_Report06042013 .pdf, accessed Dec. 2013. Matthew DeLuca, "'Absolutely Staggering': Dozens Injured in Connecticut Train Crash," NBC News, May 18, 2013, http://usnews .nbcnews.com/_news/2013/05/18/18340811-absolutely-staggering-dozens-injured-in-connecticut-train-crash, accessed Dec. 2013.

24 Jeff Plungis and Angela Greiling Keane, "Passenger Trains Bump Along with Fixes Outrunning Funding," *Bloomberg*, May 23, 2013, http://www.bloomberg .com/news/2013-05-23/trains-bump-along-with-repair-needs-outrunning-finances.html, accessed Dec. 2013.

25 Jessica Rawlinson, "Rail Travel Worldwide," Report, Mintel Group, Dec. 2011, http://www.mintel.com, accessed Dec. 2013 via Harvard University Library.

26 Robert Puentes, Adie Tomer, and Joseph Kane, "A New Alignment: Strengthening America's Commitment to Passenger Rail," Report, Brookings Institution Project on State and Metropolitan Innovation, March 2013, p. 12, http://www .brookings.edu/~/media/research/files/reports/2013/03/01%20passenger%20 rail%20puentes%20tomer/passenger%20rail%20puentes%20tomer.pdf, accessed Aug. 2014.

27 Ibid.

28 Lauren Seta, "Industry Report: Public Transportation in the US," *IBISWorld*,

June 2013, http://www.ibisworld.com, accessed Dec. 2013 via Harvard University Library. American Public Transportation Association, "Ridership by Mode and Quarter," Report, June 2013, http://www.apta.com/resources/statistics/Docu ments/APTA-Ridership-by-Mode-and-Quarter-1990-Present.xls, accessed Sept. 2013. Idem, "Transit Ridership Report: First Quarter 2013," Report, May 2013, http://www.apta.com/resources/statistics/Documents/Ridership/2013-q1-rid ership-APTA.pdf, accessed Sept. 2013. Metropolitan Transportation Authority, "The MTA Network," http://web.mta.info/mta/network.htm, accessed Dec. 2013.

29 American Public Transportation Association, "Public Transportation Investment Background Data," Report, July 2013, http://www.apta.com/resources/
reportsandpublications/Documents/Public-Transportation-Investment-
Background-Data.pdf, accessed Sept. 2013.

30 Barrie Stevens, Pierre-Alain Schieb, and Anita Gibson, *Strategic Transport Infrastructure Needs to 2030: Main Findings*, Report, Organisation for Economic Co-operation and Development, International Futures Program, 2011, p. 59, http://www.oecd.org/futures/infrastructureto2030/49094448.pdf, accessed Aug. 2014.

31 Rawlinson, "Rail Travel Worldwide."

32 Ibid.

33 Michael Renner and Gary Gardner, "Global Competitiveness in the Rail and Transit Industry," Worldwatch Institute, Sept. 2010, p. 27, http://www.world watch.org/system/files/GlobalCompetitiveness-Rail.pdf, accessed Dec. 2013.

34 Statista, "Rail Industry Dossier 2012," pp. 8, 17. Marcy Lowe et al., "U.S. Manufac-
ture of Rail Vehicles for Intercity Passenger Rail and Urban Transit: A Value Chain Analysis," Center on Globalization Governance and Competitiveness, June 2010, pp. 17, 19, http://www.cggc.duke.edu/pdfs/U.S._Manufacture_of_Rail_Vehicles_ for_Intercity_Passenger_Rail_and_Urban_Transit.pdf, accessed Dec. 2013.

35 Lowe et al., "U.S, Manufacture of Rail Vehicles," pp. 22, 27, 35–36. Renner and Gardner, "Global Competitiveness," p. 10.

36 Mark A. Wurpel, "Upgrading the Amtrak Keystone Corridor," Paper, American Railway Engineering & Maintenance of Way Association 2005 Annual Conference & Exposition, Chicago, Sept. 25–28, 2005, p. 5.

37 Ibid., pp. 7–8.

38 Ron Zeitz, "Unlocking the Keystone's Potential," Community Transportation Association, p. 30, http://web1.ctaa.org/webmodules/webarticles/articlefiles/ Unlocking_Keystone_Potential.pdf, accessed Aug. 2014.

39 Yonah Freemark, "Learning from the Keystone Corridor," *Transport Politic*, Sept. 28, 2009, http://www.thetransportpolitic.com/2009/09/28/learning-from-the-
keystone-corridor/, accessed Dec. 2013. National Railroad Passenger Corpora-
tion, "Amtrak Ridership Growth Continues in FY 2013," Press Release, April 2013, http://www.amtrak.com/ccurl/178/1001/Amtrak-Ridership-Growth-First-Six-
Months-%20FY2013-ATK-13–031.pdf, accessed Dec. 2013. Pennsylvania Depart-
ment of Transportation, *Pennsylvania Public Transportation Annual Performance Report, Fiscal Year 2007–08*, April 2009, p. 127, ftp://ftp.dot.state.pa.us/public/

bureaus/PublicTransportation/GeneralInformation/BPT%20Annual%20 Report%20FINAL%202007-08_4%2028%2009.pdf, accessed Aug. 2014. Federal Railroad Administration, "Obligated High Speed Intercity Passenger Rail Funding by Region," http://www.fra.dot.gov/Page/P0554, accessed Dec. 2013. Paul Nussbaum, "Faster Trains Coming to 'Keystone Corridor,'" *Philadelphia Inquirer*, July 22, 2011, http://articles.philly.com/2011-07-22/news/29802212_1_amtrak-trains-faster-trains-keystone-corridor, accessed Aug. 2014.

40 Virginia Avenue Tunnel, Parsons Brinkerhoff, "The Virginia Avenue Tunnel Project," http://www.virginiaavenuetunnel.com, accessed Dec. 2013. National Gateway, CSX Corporation, "Growing Demand," 2009, http://www.national gateway.org/background/growing-demand, accessed Aug. 2014. Idem, "Environmental Benefits," 2009, http://www.nationalgateway.org/benefits/environment, accessed Aug. 2014.

3. Up in the Air

1 U.S. Department of Transportation, Federal Aviation Administration, "The Economic Impact of Civil Aviation on the U.S. Economy," Aug. 2011.

2 World Economic Forum,"Travel and Tourism Competitiveness 2013," http://www3.weforum.org/docs/WEF_TT_Competitiveness_Report_2013.pdf, accessed Jan. 2014.

3 American Society of Civil Engineers, "Failure to Act: The Impact of Current Infrastructure Investment on America's Economic Future," Report, 2013, http://www.asce.org/uploadedFiles/Infrastructure/Failure_to_Act/Failure_to_Act_Report.pdf, accessed Jan. 2014.

4 Bureau of Transportation Statistics, Research and Innovative Technology Administration, "Historical Air Traffic Statistics, Annual 1954-1980," http://www.rita.dot.gov/bts/sites/rita.dot.gov.bts/files/subject_areas/airline_infor mation/air_carrier_traffic_statistics/airtraffic/annual/1954_1980.html, accessed Aug. 2014. Airlines for America, "Annual Results: US Airlines," http://www.airlines.org/data/annual-results-u-s-airlines-2/, accessed Dec. 2013. Steven Morrison and Clifford Winston, *The Economic Effects of Airline Deregulation* (Washington, D.C.: Brookings Institution Press, 1986).

5 Global Air Cargo Advisory Group, "Role of the Air Cargo Industry," http://www.gacag.org/gacag/Role.asp, accessed Jan. 2014. Center for Aviation, "World Rankings 2010: Hong Kong Eclipses Memphis as the World's Busiest Cargo Hub," http://centreforaviation.com/analysis/world-airport-rankings-2010-hong-kong-eclipses-memphis-as-the-worlds-busiest-cargo-hub-47887, accessed Jan. 2014. Bureau of Transportation Statistics, "Freight Transportation: Global Highlights," 2010, http://www.rita.dot.gov/bts/sites/rita.dot.gov.bts/files/pub lications/freight_transportation/pdf/entire.pdf, accessed Jan. 2014. TranStats Database, Bureau of Transportation Statistics, Research & Innovative Technology Administration, "Air Carrier Statistics (Form 41 Traffic)," 2014, http://www.transtats.bts.gov/, accessed Aug. 2014.

6 "Companies List," Aircraft Engine and Engine Parts Manufacturing Industry, *Hoovers Industry Analysis*, Dun & Broadstreet, 2014, http://subscriber.hoovers .com/H/industry360/companiesList.html?industryId=1802, accessed Jan. 2015.

7 Gwyn Topham, "Battle for the Future of the Skies," *Guardian*, Dec. 29, 2013, http://www.theguardian.com/business/2013/dec/29/boeing-787-dreamliner-airbus-a380-battle-for-skies, accessed Jan. 2014. Boeing & Company, "Boeing 787 Dreamliner Provides New Solutions for Airlines, Passengers," http:// www.boeing.com/boeing/commercial/787family/background.page?, accessed Dec. 2013. "Deamliner: Inside the World's Most Anticipated Airplane," CNBC, http://www.cnbc.com/id/43925643, accessed Dec. 2013. Associated Press, "Boeing 787 Dreamliner Suffers ANOTHER Battery Scare as Component Starts Smoking on Plane in Japan," *Daily Mail UK*, Jan. 14, 2014, http://www.dailymail .co.uk/news/article-2539564/Smoking-Boeing-787-Dreamliner-caused-bat tery-malfunction.html, accessed Jan. 2014.

8 Airlines for America, "Annual Results: US Airlines," http://www.airlines.org/ data/annual-results-u-s-airlines-2/, accessed Dec. 2013.

9 Associated Press, "Southwest Pilots Who Landed at the Wrong Airport Say Runway Lights Confused Them," CBS News, Jan. 17, 2014, http://www.cbsnews .com/news/southwest-pilots-who-landed-at-wrong-airport-say-runway-lights-confused-them/, accessed Jan. 2014.

10 Susan Carey and Andy Pasztor, "Report Faults Rollout of Air-Traffic-Control Upgrade," *Wall Street Journal*, Sept. 24, 2014.

11 Michael Ball et al., "Total Impact Delay Study: A Comprehensive Assessment of the Costs and Impacts of Flight Delay," Final Report, National Center of Excellence for Aviation Operations Research (NEXTOR), Oct. 2010, p. 14, http://www.isr.umd.edu/NEXTOR/pubs/TDI_Report_Final_10_18_10_V3.pdf.

12 U.S. Department of Transportation, Bureau of Transportation Statistics, "Understanding the Reporting of Causes of Flight Delays and Cancellations," 2014, http://www.rita.dot.gov/bts/help/aviation/html/understanding.html, accessed July 20.

13 Mike Bettes, "American Airlines Presents," *Morning Rush*, Weather Channel, Nov. 27, 2013, http://www.weather.com/video/american-airlines-presents-41674, accessed Dec. 2013. Stephanie Abrams, "New Total Turbulence Technology," *Morning Rush*, Weather Channel, Nov. 27, 2013, http://www.weather.com/ video/new-total-turbulence-technology-41675, accessed Dec. 2013. Omar Villafranca, "American Airlines Unveils New System for Pilots to Avoid Turbulence," *NBC News*, Dec. 11, 2013, http://www.nbcdfw.com/news/local/American-Air lines-Unveils-New-Systems-for-Pilots-to-Avoid-Turbulence-233568731.html, accessed Dec. 2013.

14 Federal Aviation Administration, "NextGen Implementation Plan," March 2011, p. 22.

15 John D Kasarda and Greg Lindsay, *Aerotropolis: The Way We'll Live Next* (New York: Farrar, Straus and Giroux, 2011).

16 Eno Center for Transportation, "Addressing Future Capacity Needs in the US Aviation System," 2013.

17 Airlines for America, "Airports Q&A." http://www.airlines.org/Pages/Airports-QA.aspx, accessed Jan. 2014.

18 Eamonn Fingleton, "The World's Best Airports—And the Only US Airport to Make the Top 30," *Forbes*, April 16, 2013, http://www.forbes.com/sites/eamonn fingleton/2013/04/16/the-worlds-best-airports-and-the-only-u-s-airport-to-make-the-top-30/, accessed Dec. 2013.

19 Airports Council International, "Passenger Facility Charges," http://www.aci na.org/sites/default/files/passenger_facility_charges_fact_sheet.pdf, accessed Dec. 2013. Eno Center for Transportation, "Addressing Future Capacity Needs."

20 Kenneth J. Button, "The Taxation of Air Transportation," Paper, School of Public Policy, George Mason University, April 2005, pp. A1–A4, http://www.gmu policy.net/transport2003/airlinetaxation.pdf, accessed Jan. 2014. Office of Policy, International Affairs, and Environment, Federal Aviation Administration, "Current Aviation Excise Tax Structure," Jan. 2013, http://www.faa.gov/ about/office_org/headquarters_offices/apl/aatf/media/Excise_Tax_Struc ture_Calendar_2013.pdf, accessed Jan. 2014. Office of Policy, International Affairs, and Environment, Federal Aviation Administration, "Airport and Airway Trust Fund (AATF) Fact Sheet," Presentation, June 2013, http://www. faa.gov/about/office_org/headquarters_offices/apl/aatf/media/AATF_Fact_ Sheet.pdf., accessed Jan. 2014. Airlines for America, "Government-Imposed Taxes on Air Transportation," 2014, http://www.airlines.org/data/govern ment-imposed-taxes-on-air-transportation/, accessed Oct. 2014. Airlines for America, "The Case for a U.S. National Airline Policy," Feb. 2013, http://www. slideshare.net/a4amediarelations/case-for-a-us-national-airline-policy, accessed Dec. 2013.

21 Eno Center for Transportation, seminar summary of "Lessons Learned from the Chicago O'Hare Modernization Program," Oct. 27, 2011.

22 Jon Hilkevitch, "Chicago, Airlines Sign Deal for One More O'Hare Runway," *Chicago Tribune, March 14, 2011,* http://articles.chicagotribune.com/2011–03–14/ news/ct-met-ohare-deal-20110314_1_o-hare-runway-lahood-airline-lawsuit, accessed Sept. 2013.

23 City of Chicago, "Mayor Emanuel Announces $7 Billion Building a New Chicago Program" Press Release, March 29, 2012, http://www.cityofchicago .org/city/en/depts/mayor/press_room/press_releases/2012/march_2012/ mayor_emanuel_announces7billionbuildinganewchicagoprogram.html, accessed Sept. 2013.

24 U.S. Census Bureau (metro areas); Federal Aviation Administration (passengers). See also Susan Carey and Cameron McWhirter, "Why Is Delta Afraid of This Tiny Airport?," *Wall Street Journal,* Dec. 17, 2013, http://online.wsj.com/ news/articles/SB10001424052702304202204579256220279763590, accessed Aug. 2014.

25 "Agreements Approved for Gary/Chicago International Airport," Aero News

Network, Jan. 29, 2014, http://www.aero-news.net/emailarticle.cfm?do=main.
textpost&id=f275916d-5212-4036-9571-ca64c5b191ad, accessed Feb. 2014.

26 U.S. Federal Aviation Administration, "Airport Privatization Pilot Program,"
Sept. 27, 2013, http://www.faa.gov/airports/airport_compliance/privatization/,
accessed Sept. 2013.

27 Richard Lincer and Adam Brenneman, "Airport Sales Take Off with LMM,"
Project Finance International Global Infrastructure Report, June 2013, pp. 25–27,
Thomson Reuters, http://pfie.reutersmedia.net/airport-sales-take-off-with-
lmm/21090612.article, accessed Dec. 2013. J. Luis Guasch, *Granting and Rene-
gotiating Infrastructure Concessions: Doing It Right* (Washington, D.C.: World
Bank, 2004), pp. 7–8.

28 Associated Press, "Midway Airport Privatization Deal Collapses," *Huffington
Post*, May 21, 2009, http://www.huffingtonpost.com/2009/04/20/midway-air
port-privatizat_n_189090.html, accessed Dec. 2013. See also John Byrne, Jeff
Coen, and Hal Dardick, "Emanuel Halts Midway Privatization Bidding," *Chicago Tribune*, Sept. 6, 2013, http://articles.chicagotribune.com/2013-09-06/
news/chi-emanuel-halts-midway-lease-talks-20130905_1_great-lakes-air
port-alliance-midway-airport-midway-advisory-panel, accessed Sept. 2013.

29 Susan Carey and Daniel Michaels, "Rise of Middle East Airlines Doesn't Fly
with U.S. Rivals: United Continental and Others Balk at Success of State-
Backed Gulf Carriers," *Wall Street Journal*, Oct. 29, 2013, http://www.wsj.com/
articles/SB10001424052702304384104579141732219208604, accessed Dec.
2013. World Economic Forum, "Travel and Tourism Competitiveness 2013,"
http://www3.weforum.org/docs/WEF_TT_Competitiveness_Report_2013
.pdf, accessed Jan. 2014. Federal Aviation Administration, *The Economic Impact
of Civil Aviation on the U.S. Economy*, Aug. 2011, http://www.faa.gov/air_traffic/
publications/media/faa_economic_impact_rpt_2011.pdf, accessed Jan. 2014.
United Nations World Tourism Organization, *Annual Report* 2012, 2013, http://
dtxtq4w60xqpw.cloudfront.net/sites/all/files/pdf/annual_report_2012.pdf,
accessed Jan. 2014, p. 8. Airlines for America, "The Case for a U.S. National Airline Policy," http://www.airlines.org/Pages/The-Case-for-a-National-Airline-
Policy.aspx, accessed Dec. 2013. "Airport Wait Times," 2012 data, U.S. Customs
and Border Protection, 2014, http://awt.cbp.gov/, accessed Aug. 2014.

4. Smart Roads Meet the Smartphone

1 "Economic Impact of Stimulating Broadband Nationally," Executive Summary, Connected Nation, Feb. 2008, http://www.connectednation.org/_docu
ments/connected_nation_eis_study_executive_summary_02212008.pdf,
accessed Jan. 2014.

2 Michigan Department of Transportation, "National Firsts," http://www.michi
gan.ogv/mdot. Gordon Sessions, *Traffic Devices: Historical Aspects Thereof*
(Washington, D.C.: Institute of Traffic Engineers, 1971).

3 Katherine Turnbull, Ken Buckeye, and Nick Thompson, "I-35W South MNPASS

Hot Lanes," Paper presented for the Transportation Research Board (TRB) 2013 Annual Meeting, p. 3.

4 Michael Janson and David Levinson, "HOT or Not: Driver Elasticity to Price on the MnPASS HOT Lanes," Nexus Research Group, Department of Civil Engineering, University of Minnesota, Sept. 2013, http://nexus.umn.edu/Papers/HOTorNOT.pdf, p. 40.

5 Shira Ovide, "Tapping 'Big Data' to Fill Potholes," *Wall Street Journal*, June 12, 2012, http://online.wsj.com/news/articles/SB10001424052702303444204577 460552615646874?mg=ren064-wsj, accessed June 2, 2014.

6 "Swedish Study Shows Digital Billboards Distract Drivers," Scenic America, 2012, http://www.scenic.org/billboards-a-sign-control/digital-billboards/swedish-digital-billboard-safety-study, accessed July 2014. William A. Perez et al., "Driver Visual Behavior in the Presence of Commercial Electronic Variable Message Signs (CEVMS)," Federal Highway Administration Report, Sept. 2012, http://www.fhwa.dot.gov/real_estate/practitioners/oac/visual_behavior_report/final/cevmsfinal.pdf, accessed July 2014.

7 Steven H. Bayless et al., "Connected Vehicle Insights: Trends in Roadway Domain Active Sensing," ITS America, Technology Scan Series 2012–14, p. 2.

8 "U.S. Vehicle Sales Market Share by Company, 1961–2013," *Wards Auto*, Jan. 26, 2014.

9 Brad Berman, "History of Hybrid Cars," June 2011, http://www.hybridcars.com/history-of-hybrid-vehicles/, accessed Sept. 2014.

10 Bayless et al., "Connected Vehicle Insights."

11 Neil Winton, "Cars That Drive Themselves Will Have Some Awesome Consequences," *Forbes*, Nov. 8, 2013, http://www.forbes.com/sites/neilwinton/2013/11/08/cars-that-drive-themselves-will-have-some-awesome-consequences/print/, accessed Nov. 2013.

12 Heather Kelly, "Driverless Car Tech Gets Serious at CES," CNN, Jan. 9, 2014, http://edition.cnn.com/2014/01/09/tech/innovation/self-driving-cars-ces/, accessed Jan. 2014.

13 David Undercoffler, "Mercedes-Benz Reveals Recent Test of Self-Driving Car," *Los Angeles Times*, Sept. 10, 2013, http://www.latimes.com/business/autos/la-fi-hy-autos-mercedes-autonomous-car-20130909,0,7249138.print.story, accessed Dec. 2013.

14 Dan Neil, "Driverless Cars for the Road Ahead," *Wall Street Journal*, Sept. 27, 2013, http://online.wsj.com/news/articles/SB10001424127887323808204579085271065923340, accessed Nov. 2013.

15 Susan Shaheen, Mark Mallery, and Karla Kingsley, "Personal Vehicle Sharing Services in North America," *Research in Transportation Business and Management* 3 (2012): 73.

16 M. G. Seigler, "Uber CEO: I Think I've Got 20,000 Years of Jail Time in Front of Me," *TechCrunch*, May 25, 2011, http://techcrunch.com/2011/05/25/uber-airbnb-jail-time/, accessed July 2014.

17 Federal Highway Administration, *Environmental Justice Emerging Trends and*

Best Practices Guidebook, Dec. 2011, sec. 1, http://www.fhwa.dot.gov/environ ment/environmental_justice/resources/guidebook/guidebook01.cfm, accessed July 2014. More-recent analysis from the Economic Policy Institute puts the figure at closer to 11 percent for a family with two parents and two children. Elise Gould et al., "What Families Need to Get By," Report, Economic Policy Institute, July 3, 2014, http://www.epi.org/publication/ib368-basic-family-budgets/, accessed July 2014.

18 Anya Kamenetz, "Could You Save Money by Losing a Car?," *Chicago Tribune*, June 10, 2014, http://articles.chicagotribune.com/2014–06–10/business/sns-201406101930–tms–savingsgctnzy-a20140610–20140610_1_car-ownership-relayrides-zipcar, accessed July 2014.

19 Callum Borchers, "Firm Tries to Make a Hub Parking Space App Fit In," *Boston Globe*, July 11, 2014.

20 Steven Bayless and Radha Needlakantan, "Smart Parking and the Connected Consumer: Opportunities for Facility Operators and Municipalities," Intelligent Transportation Society of America (ITS America), Dec. 2012, pp. 3–4. Streetline, "Becoming a Smart City: Why Cities Choose Smart Parking Solutions from Streetline," http://www.streetline.com, accessed Nov. 2013.

21 Parking Panda, "Find and Reserve Parking: Featured Cities," http://www.park ingpanda.com, accessed Nov. 2013.

22 Parkmobile, "About Parkmobile," http://www.parkmobile.com, accessed Nov. 2013.

23 Javier Gozalvez, "Wireless Connections Surpass 6 Billion Mark," *IEEE Vehicular Technology Magazine*, Dec. 2012, p. 13. Upcar Varshney, "Mobile and Wireless Technologies: 4G Wireless Networks," *IT Pro*, Sept./Oct. 2012, p. 34.

24 "LTE: The Future of Mobile Broadband Technology," Verizon Wireless White Paper, 2009, pp. 3, 15.

25 Federal Communications Commission, "700 MHz Spectrum," http://www.fcc .gov/encyclopedia/700-mhz-spectrum, accessed Dec. 2013.

26 Kenneth Button and Roger Stough, *Telecommunications, Transportation, and Location* (Northampton, Mass.: Edward Elgar, 2006), p. 137.

27 "1991–1999: Embracing Technology," UPS, About UPS, http://www.ups.com/con tent/corp/about/history/1999.html?WT.svl=SubNav, accessed Nov. 2013. "Federal Express Timeline," Federal Express, About Federal Express, http://about.van.Fed eral Express.com/article/Federal Express-timeline, accessed Nov. 2013. Rayond Fisman and Tim Sullivan, "How GPS Transformed Trucking and Made the Open Road a Lot Less Open," *Wall Street Journal*, Oct. 23, 2013, http://blogs.wsj.com/ atwork/2013/10/23/how-gps-transformed-trucking-and-made-the-open-road-a-lot-less-open/tab/print/?KEYWORDS=trucking, accessed Nov. 2013.

28 Government Accountability Office, *Intelligent Transportation Systems: Improved DOT Collaboration and Communication Could Enhance the Use of Technology to Manage Congestion*, Report GAO-12–308, March 2012, http://www.gao.gov/ assets/590/589430.pdf, accessed Nov. 2013. James Bunch et al., *Intelligent Transportation Systems Benefits, Costs, Deployment, and Lessons Learned Desk Reference: 2011 Update*, Report, U.S. Department of Transportation, Research

and Innovative Technology Administration, Sept. 2011, p. 21, http://www.itskr
.its.dot.gov/its/benecost.nsf/files/BCLLDepl2011Update/$File/Ben_Cost_
Less_Depl_2011%20Update.pdf, accessed Nov. 2013.

29 "Sizing the US and North American Intelligent Transportation Systems Market," *Daily News*, Oct. 16–20, 2011, http://www.itsinternational.us.com, accessed
Aug. 2013.

30 Gozalvez, "Wireless Connections Surpass 6 Billion Mark," p. 13.

31 International Telecommunication Union, *Measuring the Information Society*,
Report, Oct. 2013, p. 24, http://www.itu.int/en/ITU-D/Statistics/Documents/
publications/mis2013/MIS2013_without_Annex_4.pdf.

32 *Measuring the Information Society*, Report, International Telecommunication
Union, 2013, p. 44, http://www.itu.int/en/ITU-D/Statistics/Documents/publi
cations/mis2012/MIS2012_without_Annex_4.pdf, accessed Nov. 2013.

33 Richard Bennett, "No Country for Slow Broadband," *New York Times*, June 15, 2013.

34 American National Standards Institute, "Standards Activities Overview,"
http://www.ansi.org/standards_activities/overview/overview.aspx?menuid=3,
accessed Nov. 2013.

35 Preston Marshall, *The Reallocation Imperative: A New Vision for Spectrum Policy*, Report, Communications and Society Program, Aspen Institute, 2012, p. 18,
http://www.aspeninstitute.org/sites/default/files/content/docs/cands/The_
Reallocation_Imperative_A_New_Vision_For_Spectrum_Policy.pdf, accessed
Nov. 2013.

36 Ryan Knutson, "Video Boom Forces Verizon to Upgrade," *Wall Street Journal*,
Dec. 16, 2013, p. B3.

37 "NTSB Calls for Wireless Technology to Let All Vehicles 'Talk' to Each
Other," NBC News, Jan. 6, 2014, http://usnews.nbcnews.com/_news/2013
/07/23/19643634-ntsb-calls-for-wireless-technology-to-let-all-vehicles-talk-
to-each-other, accessed Jan. 2014.

5. Rethinking Cities

1 Metropolitan Planning Council, "Immeasurable Loss: Modernizing Lake Michigan Water Use," *Metropolitan Planning Council Report*, May 2013. Dan Mihalo-
poulos, "City Inaugurates Costly Plan to Replace Aged Water Mains," *New York
Times*, Dec. 17, 2011, http://www.nytimes.com/2011/12/18/us/chicago-inaugu
rates-costly-plan-to-replace-aged-water-mains.html?pagewanted=all&_r=0,
accessed Sept. 2013.

2 United Nations Department of Economic and Social Affairs, "File 21: Annual
Percentage of Population at Mid-Year Residing in Urban Areas by Major Area,
Region and Country, 1950–2050," *World Urbanization Prospects*, 2014 Revision,
http://esa.un.org/unpd/wup/CD-ROM/WUP2014_XLS_CD_FILES/WUP2014-
F21-Proportion_Urban_Annual.xls, accessed July 2014.

3 Conor Dougherty and Robbie Whelan, "Cities Outpace Suburban Growth," *Wall
Street Journal*, June 28, 2012, http://online.wsj.com/news/articles/SB10001424

052702304830704577493032619987956, accessed July 2014. Robin Lichenko, "Growth and Change in U.S. Cities and Suburbs," *Growth and Change* 32 (Summer 2001): 328. William Frey, "Population Growth in North America since 1980: Putting the Volatile 2000s in Perspective," Metropolitan Policy Program, Brookings Institution, March 2012, p. 9, http://www.brookings.edu/~/media/research/files/papers/2012/3/20%20population%20frey/0320_population_frey.pdf, accessed July 2014.

4   William Frey, "Will This Be the Decade of Big City Growth?," Brookings Institution, May 23, 2014, http://www.brookings.edu/research/opinions/2014/05/23-decade-of-big-city-growth-frey, accessed July 2014.

5   Jane Jacobs, *The Death and Life of Great American Cities* (New York: Random House, 1961).

6   Edwards Glaeser, *Triumph of the City: How Our Greatest Invention Makes Us Richer, Smarter, Greener, Healthier, and Happier* (New York: Penguin Press, 2011).

7   Daniel Chatman and Robert Nolan, "Transit Service, Physical Agglomeration and Productivity in US Metropolitan Areas," *Urban Studies*, online Aug. 1, 2013, in print vol. 51, no. 5 (March 2014): 917–37. See also http://www.citylab.com/work/2013/08/public-transit-worth-way-more-city-you-think/6532/.

8   Doug Most, *The Race Underground: Boston, New York, and the Incredible Rivalry That Built America's First Subway* (New York: St. Martin's, 2014). Also see Gavin Kleespies and Katie MacDonald, "Cambridge Transportation," Harvard Square Business Association, http://www.harvardsquare.com/history/historical-sites/cambridge-transportation, accessed July 2014. Metropolitan Transit Authority, "New York City Transit—History and Chronology," http://web.mta.info/nyct/facts/ffhist.htm, accessed July 2014.

9   Tod Newcombe, "When Will the U.S. Build Another Subway?," *Governing*, Jan. 28, 2013. Glaeser, *Triumph of the City*, p. 62.

10   Richard F. Weingroff, "Busting the Trust," *Public Roads* 77, no. 1 (July/Aug. 2013), http://www.fhwa.dot.gov/publications/publicroads/13julaug/03.cfm.

11   Equality of Opportunity Project website, http://www.equality-of-opportunity.org/index.php/city-rankings/city-rankings-100, accessed June 2014.

12   Walk Score website, http://www.walkscore.com/cities-and-neighborhoods/, accessed June 2014. Also reported in Madeline Stone, "The US Cities with the Best Public Transportation Systems," *Business Insider*, Jan. 30, 2014, http://www.businessinsider.com/cities-with-best-public-transportation-systems-2014-1, accessed June 2014.

13   U.S. Census Bureau, American Community Survey Database, 2009, http://www.census.gov/acs/www/. Also in Megan Cotrell, "Second City or Dead Last? Income Apartheid in Chicago," *Chicago Reporter*, Feb. 28, 2011, http://www.chicagonow.com/chicago-muckrakers/2011/02/second-city-or-dead-last-income-apartheid-in-chicago/, accessed July 2014. Jackeline Hwang and Robert J. Sampson, "Divergent Pathways of Gentrification: Racial Inequality and the Social Order of Renewal in Chicago Neighborhoods," *American Sociological Review* 79, no. 4 (2014): 726–51.

14   Dukakis Center for Urban and Regional Policy, "The Toll of Transportation Executive Summary," *Northeastern University*, July 2013, http://www.north

eastern.edu/dukakiscenter/wp-content/uploads/The-Toll-of-Transportation-Executive-Summary-final.pdf.

15 Massachusetts Department of Transportation, "FYI2014–FY2018 Transportation Capital Investment Plan: The Capital Budget in Plain English," 2014, p. 13, http://www.massdot.state.ma.us/Portals/0/docs/infoCenter/docs_materials/cip_FY14_FY18.pdf.

16 Aaron Reiss, "New York's Shadow Transit," *New Yorker*, June 27, 2014.

17 Texas A&M Transportation Institute, "Annual Urban Mobility Report 2012," http://mobility.tamu.edu/ums/, accessed Sept. 2013. Metropolitan Planning Council, "Bus Rapid Transit's Potential to Move Chicago's Economy . . . and Its People," *Metropolitan Planning Council Report*, Aug. 2013. Adie Tomer, "Where the Jobs Are: Employer Access to Labor by Transit," Brookings Institution, July 11, 2012, http://www.brookings.edu/research/papers/2012/07/11-transit-jobs-tomer, accessed July 2014.

18 John Greenfield, "Survey: Most Chicagoans Support Bus Rapid Transit," *Streetsblog Chicago*, July 26, 2013, http://chi.streetsblog.org/2013/07/26/survey-most-chicagoans-support-bus-rapid-transit/, accessed Sept. 2013. BRT Chicago, "BRT Chicago Partners Overview," http://www.brtchicago.com/partners.php, accessed Sept. 2013.

19 BRT Chicago, "BRT Chicago Loop Overview," http://www.brtchicago.com/brt loop.php, accessed Sept. 2013.

20 A group of Harvard Advanced Leadership Fellows from my program toured Denver's Union Station in June 2014 and reported to me. See also Taras Grescoe, "How Denver Is Becoming the Most Advanced Transit City in the West," *Atlantic: CityLab*, June 24, 2014, http://www.citylab.com/commute/2014/06/how-denver-is-becoming-the-most-advanced-transit-city-in-the-west/373222/, accessed June 2014. Institute for Sustainable Communities, "Case Study: Denver FasTracks," 2010, http://www.iscvt.org/resources/documents/denver_fastracks.pdf, accessed June 2014. Diane Barrett, "Financing Denver Union Station," Presentation, Institute for Sustainable Communities, http://www.iscvt.org/where_we_work/usa/article/low_carbon_transportation/barrett_denver.pdf, accessed July 2014. Judy Montero, Denver city councilwoman, as quoted in RTD FasTracks, "Union Station Reopens," YouTube, published May 9, 2014, https://www.youtube.com/watch?v=ahh-VwFiON8#t=845, accessed June 2014. Don Hunt, executive director, Colorado Department of Transportation, as quoted ibid.

21 Ben Block, "In Amsterdam, the Bicycle Still Rules," WorldWatch Institute, http://www.worldwatch.org/node/6022, accessed June 2014.

22 "NHTSA Data Confirms Traffic Fatalities Increased in 2012," NHTSA Press Release, Nov. 14, 2013, http://www.nhtsa.gov/About+NHTSA/Press+Releases/NHTSA+Data+Confirms+Traffic+Fatalities+Increased+In+2012, accessed July 2014.

23 U.S. Census, "Biking to Work Increases 60 Percent over Last Decade, Census Bureau Reports," Press Release, May 28, 2014, https://www.census.gov/news room/releases/archives/american_community_survey_acs/cb14–86.html, accessed July 2014. U.S. Census, American Fact Finder, http://factfinder2.cen

sus.gov/faces/tableservices/jsf/pages/productview.xhtml?pid=ACS_12_1YR_
B08301&prodType=table, accessed July 2014.

24 Angie Schmitt, "The Rise of the North American Protected Bike Lane," *Momentum Magazine*, July 31, 2013, http://momentummag.com/features/the-rise-of-
the-north-american-protected-bike-lane/, accessed July 2014.

25 Zack Furness, *One Less Car: Bicycling and the Politics of Automobility* (Philadelphia: Temple University Press, 2010), pp. 55–59.

26 City of Chicago Department of Transportation, "Chicago Complete Streets: Federally Funded Bike Lanes in the Windy City," 2006, http://www.chicagobikes
.org/pdf/complete_streets_bikeways.pdf, accessed June 2013. City of Chicago,
"Bike 2015 Plan," http://www.bike2015plan.org/intro.html, accessed Jan. 2014.

27 Deanna Issacs, "Is the City's Shiny New Bike-Sharing Program a Dirty Deal?,"
Chicago Reader, May 9, 2012, http://www.chicagoreader.com/chicago/bike-shar
ing-contract-may-be-inside-job/Content?oid=6274245, accessed Sept. 2013.
Brian Costin, "Chicago Bike-Sharing Program Takes Taxpayers for a Ride," Illinois Policy Institute, Nov. 26, 2012, http://www.illinoispolicy.org/chicago-bike-
sharing-program-takes-taxpayers-for-a-ride/, accessed Sept. 2013. Rebel Pundit,
"Chicago Residents Go to Court over Rahm's Rental Bikes," *Breitbart*, Aug. 25,
2013,   http://www.breitbart.com/Big-Government/2013/08/25/Chicago-Resi
dents-Go-to-Court-over-Rahm-s-Rental-Bikes, accessed Sept. 2013.

28 National Highway Traffic Safety Administration Data, http://www.nhtsa
.gov/NCSA. See also Chuck Sudo, "Chicago Pedestrian Deaths Spiked in 2012,"
Chicagoist, Jan. 22, 2013, http://chicagoist.com/2013/01/22/chicago_pedes
trian_deaths_spiked_in.php, accessed Dec. 2013. See also John Greenfield, "Hit-
and-Run Van Driver Kills Senior in Englewood," *Streetsblog Chicago*, Jan. 2,
2014, http://chi.streetsblog.org/tag/fatality-tracker/page/2/, accessed July 2014.

29 John Heilemann, "Reinventing the Wheel," *Time*, Dec. 2, 2001, http://content
.time.com/time/business/article/0,8599,186660–1,00.html, accessed June 2014.
"Segway Recalls 23,500 Scooters," *CNN Money*, Sept. 14, 2006, http://money.
cnn.com/2006/09/14/news/companies/segway/index.htm, accessed June 2014.
Mark Hachman, "Segway Quietly Sold; Dealers Remain Optimistic," *PC Magazine*, Jan. 18, 2010, http://www.pcmag.com/article2/0,2817,2358173,00.asp,
accessed June 2014. "Summit Strategic Investments, LLC Acquires Segway Inc.,"
Yahoo Finance, Feb. 28, 2013, http://finance.yahoo.com/news/summit-strategic-
investments-llc-acquires-214253655.html, accessed June 2014.

30 Superpedestrian, "The Copenhagen Wheel Official Product Release," YouTube,
published Dec. 3, 2013, https://www.youtube.com/watch?v=S10GMfG2NMY,
accessed July 2014. SENSEable City Lab, Massachusetts Institute of Technology,
"The Wheel," 2009, http://SENSEable.mit.edu/copenhagenwheel/wheel.html,
accessed July 2014. Assaf Biderman, "A Note from Our Founder," *Superpedestrian
Blog*, May 1, 2014, http://copenhagenwheel.tumblr.com/post/84424714259/a-
note-from-our-founder, accessed July 2014. Curt Woodward, "Superpedestrian
Starts Selling 'Copenhagen Wheel' Electric Bike Kits," *Xconomy*, Dec. 3, 2014,
http://www.xconomy.com/boston/2013/12/03/superpedestrian-starts-selling-
copenhagen-wheel-electric-bike-kits/, accessed July 2014.

31 Benjamin Barber, *If Mayors Ruled the World: Dysfunctional Nations, Rising Cities* (New Haven: Yale University Press, 2013).

32 Rosabeth Moss Kanter and Ai-Ling Jamila Malone,"IBM and the Reinvention of High School (A): Proving the P-TECH Concept," Harvard Business School Case 314–049, Sept. 2013. Idem, "IBM and the Reinvention of High School (B): Replicating & Scaling P-TECH and Partners," Harvard Business School Case 314–050, Sept. 2013.

6. The Will and the Wallet

1 American Association of Port Authorities, "NAFTA Region Container Traffic: 2012 Port Ranking by TEUs," 2012, http://aapa.files.cms-plus.com/Statistics/NAFTA%20REGION%20CONTAINER%20TRAFFIC%20PORT%20RANKING%202012.pdf.

2 "Fact Sheet: Modernizing and Investing in America's Ports and Infrastructure," Press Release, White House, Nov. 8, 2013, http://www.whitehouse.gov/the-press-office/2013/11/08/fact-sheet-modernizing-and-investing-america-s-ports-and-infrastructure. American Association of Port Authorities, "U.S. Public Port Facts," July 2008, http://www.aapa-ports.org/files/pdfs/facts.pdf. Idem, "Infrastructure Improvements," http://www.aapa-ports.org/Industry/content.cfm?ItemNumber=1025&navItemNumber=1029, accessed Jan. 28, 2013. American Society of Civil Engineers, *Failure to Act: The Economic Impact of Current Investment Trends in Airports, Inland Waterways, and Marine Ports Infrastructure*, Report, 2012, p. 11, http://www.asce.org/uploadedFiles/Infrastructure/Failure_to_Act/Failure%20To%20Act%20Ports%20Economic%20Report.pdf.

3 Mimi Whitefield, "Arrival of Giant Cranes Ushers In New Era at PortMiami," *Miami Herald,* Oct. 8, 2013, http://www.miamiherald.com/2013/10/07/3676145/arrival-of-giant-cranes-ushers.html.

4 Port of Miami Tunnel, Florida Department of Transportation, "Project Overview," 2011, http://www.portofmiamitunnel.com/project-overview/project-overview-1/.

5 Jim Rush, "State-of-the-Art," *Tunnel Business Magazine*, June 4, 2013, http://tunnelingonline.com/state-of-the-art, accessed Dec. 2013. Jeffrey A. Parker, "Port of Miami Tunnel Breaks New Ground for Greenfield P3 Project in the U.S.," *Public Works Financing*, no. 243 (Nov. 2009), http://www.pwfinance.net, accessed Dec. 2013.

6 Manny Diaz, *Miami Transformed: Rebuilding America One Neighborhood, One City at a Time* (Philadelphia: University of Pennsylvania Press, 2013), p. 162.

7 "TIFIA Credit Program Overview," Presentation, TIFIA Joint Program Office, Office of Innovative Program Delivery, Federal Highway Administration, U.S. Department of Transportation, Dec. 2013, p. 1, http://www.fhwa.dot.gov/ipd/pdfs/tifia/bkgrnd_slides_december_2013.pdf.

8 "POMT Reaches Financial Close," *Project Finance International*, no. 419 (Oct. 21, 2009), http://www.pfie.com.

9   Scott Blake, "Unprecedented Access to Buoy PortMiami," *Miami Today*, Dec. 24, 2013, http://www.miamitodaynews.com/2013/12/24/unprecedented-access-buoy-portmiami/, accessed May 2014.

10 Alfonso Chardy, "Governor Dedicates New PortMiami Tunnel," *Miami Herald*, May 19, 2014, http://www.miamiherald.com/2014/05/19/4124977/governor-dedicates-new-portmiami.html, accessed July 2014.

11 AASHTO Center for Excellence in Project Finance, American Association of State Highway and Transportation Officials, "Private Activity Bonds," 2013, http://www.transportation-finance.org/funding_financing/financing/bond ing_debt_instruments/municipal_public_bond_issues/private_activity_bonds .aspx, accessed Jan. 2014. Office of Innovative Program Delivery, Federal Highway Administration, U.S. Department of Transportation, "Private Activity Bonds," 2013, http://www.fhwa.dot.gov/ipd/finance/tools_programs/federal_ debt_financing/private_activity_bonds/index.htm, accessed Jan. 2014. Reuters, "Roadway Bonds Lead U.S. Municipal Sales Next Week," July 3, 2014, http:// www.reuters.com/article/2014/07/03/markets-municipals-deals-idUSL2N-0PE19Y20140703, accessed July 2014.

12 American Society of Civil Engineers, *2013 Report Card for America's Infrastructure* (Reston, Va.: American Society of Civil Engineers, 2013), http://www.infra structurereportcard.org.

13 National Surface Transportation Policy and Revenue Study Commission, *Transportation for Tomorrow: Report of the National Surface Transportation Policy and Revenue Study Commission* (Washington, D.C.: GPO, Dec. 2007), p. 6, http://transportationfortomorrow.com/final_report.

14 "Life in the Slow Lane," *Economist*, April 28, 2011, http://www.economist.com/ node/18620944.

15 Barbara Weber and Hans Wilhelm Alfen, *Infrastructure as an Asset Class: Investment Strategies, Project Finance and PPP* (Chichester, West Sussex: John Wiley, 2010), p. 55.

16 Bethany McLean, "Would You Buy a Bridge from This Man?," *CNN Money*, Oct. 2, 2007, http://europe.cnnfn.com/2007/09/17/news/international/macquarie_ infrastructure_funds.fortune/index.htm.

17 Eduardo Engel, Ronald Fischer, and Alexander Galetovic, "Public-Private Partnerships to Revamp U.S. Infrastructure," Discussion Paper, Hamilton Project, Brookings Institution, Feb. 2011, p. 10, http://www.ncppp.org/wp-content/ uploads/2013/03/PS-Feb2011-HamiltonProject.pdf.

18 Ryan Holeywell, "The Indiana Toll Road: A Model for Privatization?," *Governing*, Oct. 2011, http://www.governing.com/topics/mgmt/indiana-toll-road-model-privatization.html, accessed Dec. 2013. Ryan Dezember and Emily Glazer, "Drop in Traffic Takes Toll on Investors in Private Roads," *Wall Street Journal*, Nov. 20, 2013, http://online.wsj.com/news/articles/SB10001424052702303482 504579177890461812588. "Indiana Toll Road Operator Facing Debt Woes," *Indiana Business Journal*, June 19, 2014, http://www.ibj.com/indiana-toll-road-operator-facing-debt-woes/PARAMS/article/48208.

19 "Editorial: Chicago Parking Meter Deals [*sic*] Needs Changes," *Chicago Tribune*, June 2, 2013, http://articles.chicagotribune.com/2013–06–02/opinion/ct-edit-meters-20130602_1_chicago-parking-meters-llc-chicago-city-council-aldermen, accessed July 2014.

20 Dick Johnson, "Parking Meter Revenue Up $30M in 2012," *NBC Chicago*, May 7, 2013, http://www.nbcchicago.com/investigations/chicago-parking-meters-2012-revenue-206533221.html, accessed July 2014.

21 Hal Dardick, "Parking Meter Firm's Take from the City Rises," *Chicago Tribune*, July 3, 2014, http://articles.chicagotribune.com/2014–07–03/news/ct-chicago-parking-meters-met-0703–20140703_1_chicago-parking-meters-city-rises-free-parking, accessed July 2014.

22 Robert Bain and Lidia Polakovic, *Traffic Forecasting Risk Study Update 2005: Through Ramp-Up and Beyond*, Standard & Poor's Global Credit Portal, Aug. 25, 2005, p. 3, http://www.robbain.com/Traffic%20Forecasting%20Risk%20 2005.pdf.

23 Young Hoon Kwak, YingYi Chih, and C. William Ibbs, "Towards a Comprehensive Understanding of Public Private Partnerships for Infrastructure Development," *California Management Review* 51, no. 2 (Winter 2009): 52.

24 Jason Kelly, "KKR Raises $4 Billion for Deals in Infrastructure, Energy," *Bloomberg*, June 26, 2012, http://www.bloomberg.com/news/2012–06–26/kkr-raises-4-billion-for-deals-in-infrastructure-energy.html.

25 Richard Dobbs, Herbert Pohl, Diaan-Yi Lin, et al., *Infrastructure Productivity: How to Save $1 Trillion a Year* (New York: McKinsey Global Institute, Jan. 2013), p. 24, http://www.mckinsey.com/insights/engineering_construction/infra structure_productivity.

26 Ibid., p. 7.

27 European Investment Bank, *Financial Report 2012*, EIB Group, March 2013, pp. 26–27, http://www.eib.org/attachments/general/reports/fr2012en.pdf.

28 Brina Milikowsky, *Building America's Future: Falling Apart and Falling Behind: Transportation Infrastructure Report 2012* (Washington, D.C.: Building America's Future Educational Fund, 2012), pp. 28–29, http://www.bafuture.org/pdf/Building-Americas-Future-2012-Report.pdf.

29 Joung Lee, "Five Myths about the TIFIA Credit Program," *Eno Brief Newsletter* (Eno Center for Transportation), Nov. 5, 2012, http://www.enotrans.org/eno-brief/five-myths-about-the-tifia-credit-program.

30 Keith Miller, Kristina Costa, and Donna Cooper, *Creating a National Infrastructure Bank and Infrastructure Planning Council: How Better Planning and Financing Options Can Fix Our Infrastructure and Improve Economic Competitiveness* (Washington, D.C.: Center for American Progress, Sept. 2012), pp. 8–9.

31 Robert Puentes and Jennifer Thompson, "Banking on Infrastructure: Enhancing State Revolving Funds for Transportation," Paper, Brookings-Rockefeller Project on State and Metropolitan Innovation, Sept. 2012, p. 7.

32 Greg Hinz, "Beitler Sets Big Goals for Chicago Infrastructure Trust," *Crain's Chicago Business*, Feb. 4, 2013, http://www.chicagobusiness.com/article/20130204/

BLOGS02/130209945/beitler-sets-big-goals-for-chicago-infrastructure-trust, accessed Sept. 2013. MarySue Barrett, "Op-Ed: Chicago Needed an Investment Partner, So It's Building One Itself," *Next City*, April 18, 2013, http://nextcity.org/daily/entry/op-ed-chicago-needed-an-investment-partner-so-its-building-one-itself, accessed Sept. 2013.

33 Bill Ruthhart, "Chicago Infrastructure Trust OKs $25 Million Energy Efficiency Project," *Chicago Tribune*, Nov. 12, 2013, http://articles.chicagotribune.com/2013–11–12/news/ct-met-chicago-infrastructure-trust-1113–20131113_1_energy-efficiency-retrofit-chicago-city-council.

34 Lee, "Five Myths."

35 "Information on the Partnership to Build America Act (H.R. 2084)," Office of Congressman John Delaney (MD-6), Nov. 20, 2013, http://delaney.house.gov/information-on-congressman-delaneys-infrastructure-bill.

7: How to Move

1 "Garrett A. Morgan Technology and Transportation Education Program," Brochure, Technology Partnership Programs, Federal Highway Administration, http://www.fhwa.dot.gov/tpp/gamttep_brochure.pdf, accessed Oct. 2014. "About Us," TransTech STEM Academy, Francis L. Cardozo Senior High School, http://transtechacademy.com/about-us/, accessed Oct. 2014.

2 Robert Puentes and Bruce Katz, "To Fix America's Infrastructure, Washington Needs to Get Out of the Way," *Forbes*, May 9, 2014, http://www.forbes.com/sites/realspin/2014/05/09/to-fix-americas-infrastructure-washington-needs-to-get-out-of-the-way/, accessed Oct. 2014. "Governor O'Malley to Facilitate Mid-Atlantic Regional Workgroup on Infrastructure Investment," Press Release, Office of Governor Martin O'Malley, State of Maryland, June 27, 2014, http://www.governor.maryland.gov/blog/?p=10524, accessed Oct. 2014. The West Coast Exchange has echoes of an earlier vision for the region, given the name "Cascadia"; Rosabeth Moss Kanter, *World Class: Thriving Locally in the Global Economy* (New York: Simon and Schuster, 1995).

ACKNOWLEDGMENTS

Writing a book on such a big topic can seem a lonely task. It certainly had grueling moments, as I moved from one great place to another preoccupied with ideas, and from office to home with my laptop stuffed with digital files. But it's not lonely at all when there's a big team effort behind it. I want to start by thanking a great team: my Harvard Business School research associates Daniel Fox, Ai-Ling Malone, Kevin Rosier, and Matthew Guilford, who worked full-time on the payroll and beyond to identify facts and figures, travel with me, and conduct interviews. Dan and Ai-Ling lived with this the longest and were always cheerful, responsive, hardworking, and intelligent in person and across cyberspace, doing heroic work; Dan was a tireless and smart investigator all the way to completion. Erin Henry and Aditi Jain also contributed to the research, as did Allison Belanger. Richard Edelman provided outstanding insights into how best to talk about the issues, backed by a great Edelman team.

The work received generous financial support and valuable encouragement from Harvard Business School Dean Nitin Nohria, the Ann and Andrew Tisch Foundation (thank you, Andy), and Paul Healy, Senior Associate Dean leading the Division of Research and Faculty Development. DRFD staff, including Robyn Mukherjee and Maggie Hall, ably supported the America on the Move Summit, which was connected to the U.S. Competitiveness Project, chaired by

Professors Michael E. Porter and Jan Rivkin. Summit sessions were brilliantly facilitated by my stellar HBS colleagues Amy Edmondson, Ben Esty, Janice Hammond, Rebecca Henderson, David Moss, Michael Porter, John Quelch, Jan Rivkin, and Willy Shih. Rakesh Khurana and Stephanie Ralston Khurana provided valuable encouragement at key moments.

Experts at other parts of Harvard and MIT were helpful at early planning stages, and some of them spoke at the summit. They include Charles Ogletree at Harvard Law School; Harvard Kennedy School faculty members Akash Deep, Edward Glaeser, David Gergen, Jose Gomez-Ibanez, Jassim Jaidah, and Henry Lee; and MIT faculty Chris Caplice, Daniela Rus, and Anthony Vanky. Volpe Center staff were also very helpful at the outset: Robert Johns, Anne Aylward, Ellen Bell, Kam Chin, Michael Dinning, Ryan Harrington, Stephen Popkin, and Gary Ritter. Conversations with Governor Deval Patrick and fellow members of the Governor's Council of Economic Advisors helped shape my thinking as well. I also learned from serving on the Milstein Commission on Infrastructure and the Middle Class, chaired by former U.S. Transportation Secretary Ray LaHood and former Los Angeles Mayor Antonio Villaraigosa.

Big thanks are due those who toil in the transportation sector every day, who do outstanding work and have produced outstanding analyses. Kudos to experts at nearly a hundred associations and think tanks, which are too numerous to thank individually but show up in the references in the notes and were listed in a catalog my team compiled, which is available on the HBS website. But I do want to single out for thanks Robert Atkinson of the Information Technology and Innovation Foundation, Robert Puentes and Bruce Katz of the Brookings Institution, Joshua Schank of the Eno Center for Transportation, and Jimmy Hexter formerly of McKinsey.

The honorable and esteemed Rodney Slater was available every step of the way, and spoke eloquently at the summit and subsequent events. Douglas Foy was generous with time and insights. Others who lent their wisdom and experience to the summit or on field visits include Charlie Bolden, Virginia Rometty, Stanley Litow, Scott Griffith, David Kenny, Mark Gildersleeve, Mark Miller, Desmond Keany, Lowell McAdam, Jim Gerace, Brian Higgins, Stephen Koch, MarySue Barrett, Gabe Klein, Manny Diaz, Jose Abreu, Chris Hodgkins, Joseph Aiello, David Kiley, Chris Beall, Beverly Scott, Anita Hairston, Michael Lewis, and Ann Drake. Others are named or credited in the pages of this book.

I also want to thank the unsung heroes on numerous staffs who, as assistants and schedulers, helped facilitate connections with national leaders such as U.S. Secretary of Transportation Anthony Foxx (and also his mother, Laura Foxx); U.S. Senators Chris Coons, Edward Markey, Barbara Mikulski, and Elizabeth Warren; and U.S. Representatives Earl Blumenauer and Tom Rice. Also staffs for executives such as Jeffrey Immelt, Karen Mills, Patrick Gallagher, Richard Anderson, Michael Ward, Wick Moorman, Ben Minicucci, David Seigel, Charlie Bolden, Mary Barra, Bill Graves, Karen Freeman-Wilson, Jerry Storch, Laura Sen, Chris Sultemeir, William Logue, Richard Trumka, and Tom Donohue.

Harvard Advanced Leadership Fellows for 2014 discussed the issues in this book with great insight, bringing their collective wisdom as leaders at the top of their fields now interested in leading social change. I am especially grateful to Bruce Cohen, Susan Giannino, Carol Goss, Carol Johnson, Guy Rollnick, Torsten Thiele, and Waide Warner for lending wide-ranging global and national expertise.

Readers of portions of the book in draft were immensely helpful in identifying gaps, suggesting sources, and correcting matters of fact, although, of course, the responsibility for the final contents of this book is solely mine. I especially want to thank my astute Harvard colleagues Rakesh Khurana, Ryan Raffaelli, Jan Rivkin, and Willy Shih. Other readers to whom I'm very grateful include Rob Atkinson, Dan Fox, Doug Foy, Jane Garvey, Marc Hoecker, David Kenny, David Kiley, Rodney Slater, Maxeme Tuchman, Michael Ward, and Waide Warner. I wish I could have included everything they added. W. W.

Norton president/publisher/editor Drake McFeely and editor Jeffrey Shreve were just what editors should be: discerning and tough. I benefited from their structural and substantive comments.

Barry Stein was patient and always interested. Matthew Moss Kanter Stein reminded me numerous times of Kanter's Law: that everything looks like a failure in the middle. Alison Lily Stein let me borrow her copy of *Oh, the Places You'll Go!* by Dr. Seuss, where we read about getting stuck waiting for buses, trains, and nearly anything else, and the spirit to keep moving to a better place. Natalie Elyse Stein joined me in reading *The Little Engine That Could*. Three-peat Stein kept us all in the Waiting Place, especially Melissa Kamin Stein, who bore the burden. Then Three-peat arrived under the name of Jacob Harrison Stein as this book was being completed. All is well.

—Boston, Cambridge, Edgartown, and Miami, 2014

INDEX

Abreu, Jose, 23, 226–28, 234–35, 240, 271

Acela Express, 29, 59, 62, 70, 200

Advanced Leadership Initiative, 2

Aerostar Airport Holdings, 110

Aerotech, 98

aerotropolis, 101

Aiello, Joseph, 233–34, 237, 250

Airbus, 80, 98, 115

air cargo, 74, 78–80, 104, 110, 114

Airglades Airport, 110

Airline Deregulation Act (1978), 78

Airlines for America, 116

air pollution, 1, 9–10, 13, 16, 41, 42, 74, 75, 76, 86, 87, 95, 103, 116, 130, 147, 148, 167, 176, 276

   see also carbon emissions

Airport and Airway Trust Fund, 103–4

Airport Area Task Force, 108

airports, 73–74, 75, 78, 79, 81, 101–13, 247

   see also specific airports

Airports Council International—North America, 102–3

Airspace Optimization and Aircraft Technology, 76

air traffic control, 73, 76, 78, 83–84, 86, 87, 88, 89, 93, 100–101, 103–4, 271–72, 277–78

air travel, 73–117

   aircraft for, 33–34, 80–89, 91, 92, 98, 100, 115

   airliner fleets of, 80, 83, 87, 88–89

   airport privatization in, 109–11, 247, 270

   airports for, 73–74, 75, 78, 79, 81, 101–13, 247; *see also specific airports*

   air traffic control for, 73, 76, 78, 83–84, 86, 87, 88, 89, 93, 100–101, 103–4, 271–72, 277–78

   arrivals and departures of, 93–94, 100–101

   baggage on, 111, 115

   bankruptcies in, 73, 82–83

   bureaucracy in, 90–91, 115–16

   business class on, 78, 81, 82

   carbon emissions of, 13, 42, 74, 75, 76, 86, 95, 103, 116

   cockpits in, 90–91, 98–99

   communications for, 25, 76, 87–93, 102–3, 117

   competition in, 74–75, 78, 80, 81, 84–86, 111–14

   computerization of, 78, 79, 83, 87–100

   congressional oversight of, 74, 89, 116–17

air travel (*continued*)
    customer satisfaction in, 73, 74, 76, 82,
      92, 102
    decision-making in, 88, 93–94, 99, 100,
      103, 117
    delays in, 1, 10, 11, 24, 73, 74, 75, 85,
      93–96, 99, 100–102, 105–6, 108, 109,
      116, 117
    deregulation of, 78, 83
    economic impact of, 74, 82–83, 94–95,
      101, 104–7, 112–13
    entrepreneurship in, 90–91
    environmental impact of, 73, 80, 90,
      105
    federal regulation of, 73, 76–77, 83, 84,
      89–90, 92, 93, 111–12, 116–17
    federal subsidies for, 30, 61, 84
    flight attendants in, 91–92, 100
    foreign investment in, 108, 112–13
    freight (air cargo) carried by, 74,
      78–80, 104, 110, 114
    frequent-flyer programs in, 78, 104
    fuel efficiency of, 74, 75, 80, 81–82,
      85–86, 87, 95, 99, 100, 101
    hub-and-spoke system in, 101–2, 107,
      110
    industry for, 61, 76–83, 85, 87, 104–7,
      110–11, 116, 247
    intermodal connections for, 103
    international, 73, 74, 75, 78, 79, 102,
      111–15
    investment in, 75, 87, 89, 94, 102–4,
      108, 111, 247
    jet aircraft for, 33–34
    job creation in, 74, 106, 107, 108, 109
    leadership in, 74, 75–76, 88, 90–94, 99,
      100, 103, 117
    legacy airlines in, 80, 83
    long-distance, 80–81, 114–15
    mergers of, 83, 111
    military, 33–34, 80, 83
    modernization of, 73, 115–16, 160
    navigation technology for, 25, 83–87,
      92

    noise reduction in, 76, 86, 87, 103
    nonstop flights in, 114
    passengers service of, 73, 74, 77–78,
      93–94, 99–100, 102
    pilots for, 75–76, 85, 87, 88, 90–94, 95,
      100, 116, 273, 277–78
    political aspect of, 105–7, 110
    pollution from, 13, 42, 74, 75, 76, 86, 87,
      95, 103, 116
    profitability of, 79–83, 85, 104–5, 116
    public-private partnerships in, 110–11
    radio navigation in, 83–84, 85, 86, 92
    rail connections for, 11, 24, 28, 33, 73,
      74–75, 197, 200, 201
    regional, 82, 107–9, 111, 115
    runways for, 101, 105–7, 108, 109
    safety of, 78, 81, 86, 91, 94–95, 99–100,
      116
    satellite navigation for, 84–87
    seating in, 74
    seat-miles in, 74
    security at, 81, 104–5
    shuttle service of, 59, 74
    subway connections for, 180, 191, 195
    tarmac delay penalties in, 95–96
    taxation and, 75, 103–4, 111
    technology for, 73, 74, 76, 83–85, 86,
      87–94, 100–101, 116–17, 160, 166,
      271–72
    ticket prices in, 73, 74–75, 78, 82,
      103–4
    turbulence in, 25, 73–74, 95, 99–100
    visa requirements and, 113–14
    visibility in, 85
    weather conditions for, 10, 25, 73–74,
      75, 76, 85, 87–88, 91, 92, 94–96,
      99–100, 105–6, 277–78
Alameda Corridor, 41
Alaska, 85, 221
Alaska Airlines, 83, 85
Alcatel Lucent, 135
All Aboard Florida, 70
Allegiant Air, 109
Alonzo, Joe, 50

Alstom, 63

Alta Bicycle Share, 204, 206

Alvarez, Carlos, 228, 234

Amazon, 160, 167

American Airlines (AA), 3, 75–76, 78, 80,
    81, 83, 84–85, 86, 87–89, 92, 94, 98,
    99–100, 104, 106–7, 111, 126, 160, 228,
    277

American Airlines Arena, 225–26

American Airlines Flight 1916, 86

American Association of State Highway
    and Transportation Officials, 39

American Automobile Association, 145

American Motors, 128

American National Standards Institute,
    166

American Public Transportation
    Association, 177, 178

American Recovery and Reinvestment
    Act (2009), 67, 161

American Road & Transportation
    Builders Association, 232

American Society of Civil Engineers, 75,
    214, 243

"America on the Move: Transportation
    and Infrastructure for the 21st
    Century" summit (2014), 2–3, 18, 39,
    144, 254

Amsden, Mike, 198, 199, 208

Amsterdam, 203–4

Amtrak, 29, 34, 35, 44–45, 56, 57–68,
    199–200, 201, 273

Anderson, Richard, 81–82, 104, 105

Android, 136, 141, 155, 165

antitrust legislation, 33, 46, 83

Apple Computers, 11, 90, 91–92, 155, 164

apps, 7, 22, 26, 121, 126, 127, 135, 136,
    139–54, 155, 165, 276

Arsht Performing Arts Center, 225–26

Ashland Avenue, 196–97, 207

Asia Infrastructure Investment Bank, 251

ASUR, 110

AT&T, 160

Atkinson, Robert, 138, 163, 164–65, 168

Atlanta, 10, 24, 107–9, 189–90, 267

Atlanta Aerotropolis Alliance, 108

Atlanta Regional Commission, 108

Audi, 138

auto accidents, 12, 24, 125, 127, 132–33,
    139, 162, 207, 209, 277

Automated Engine Start Stop, 42

Automatic Dependent Surveillance
    Broadcast (ADS-B), 86

automobile industry, 34, 50, 120, 122,
    128–31, 134, 143

automobiles, see cars

Auxiliary Power Units, 42

Aviation Facilities Co., 109

aviation industry, 61, 76–83, 85, 87, 104–7,
    110–11, 116, 247

Avis Budget Group, 144

AvPorts, 109

Aykac, Adnan, 115

Aylward, Anne, 137

Babcock & Brown, 233, 235, 236

Bakken crude oil, 45

Baltimore, Md., 53, 123

Baltimore & Ohio Railroad, 36

Baltimore and Potomac Railroad, 52–53

Banavar, Guruduth, 133

Barger, Dave, 77

Barra, Mary, 3, 131, 137

Barrett, MarySue, 206, 214, 253

Basulto, Eduardo, 240

Bay Area Rapid Transit (BART), 179, 180

Beitler, Steven, 253

Bell, Ellen, 137

Berkeley, Calif., 122–23

Berkshire Hathaway, 38

"Best of ITS" award, 124

bicycle lanes, 25, 207, 209

bicycle paths, 265

bicycles, 26, 68, 175, 176, 186, 201, 202–12,
    218, 265, 269–70, 276, 279

bicycle-sharing programs, 3, 203–4, 218,
    269–70, 276

bicycle stations, 206–10, 212

Biderman, Assaf, 211
BigBelly, 157–58
Big Data, 24, 96–97, 119, 121–27, 128, 140,
    155, 157, 166, 167, 192, 262, 263, 273
Big Dig, 176
Bike 2000 plan, 205
Bike 2015 plan, 205
Bikes Not Bombs, 212
Bikes USA, 205
billboards, 125–26, 215
"black boxes," 42
black-car services, 141–42
Blackstone, 249
blind pedestrians, 126–27
Bloomberg, Michael, 24, 204, 213, 216
Bloomingdale Trail, 216
Blue Cross/Blue Shield, 208
Blumenauer, Earl, 278
BMW, 138
BNSF, 38
Bob the Builder, 239
Boeing airliners, 80–81, 84–85, 87–89, 91,
    92, 98, 100
Boeing Co., 33–34, 80–81, 85
Bombardier, 63, 80
bonds, 198, 207, 236, 241–42, 253
Boston, 7, 9, 10, 60, 81, 121, 148–49, 176,
    179, 183, 184, 185, 189, 192–93, 195,
    213, 230
Bouygues Travaux Publique, 229–32, 233,
    234, 236, 240, 248, 250
Bradley International Airport, 86
Brazilian National Development Bank,
    251
Bridgeport, Conn., train crash (2013),
    9–10, 55–57
bridges, 1, 7, 10, 15, 21, 23, 50, 51, 52, 120,
    122, 172, 221, 252, 256, 264, 265
Bridj, 191–93, 205, 273
British Airways, 78, 111
broadband communications, 12, 17, 113,
    122, 125, 128, 131, 135–36, 144, 148,
    155–67, 192, 215, 272, 277
Brooklyn, 149, 205

Brookville, 63
Brown, Jerry, 24, 275–76
Buffett, Warren, 38
BUILD Act, 254
Building a New Chicago, 214
bullet trains, 27, 35, 64, 66, 70, 103
Burgess, George, 235
Bush, Jeb, 226, 227–28, 234
Bus Rapid Transit (BRT), 194–99, 206,
    207, 208, 209, 215, 218, 268
bus transportation, 10–11, 21, 22, 26, 61,
    117, 157, 159, 172, 177–78, 184, 185,
    186–87, 190–201, 206, 207, 208, 209,
    211, 215, 218, 268, 273
"Buy American" rules, 64, 129

California, 6, 32, 41, 68, 122–23, 141, 143,
    275–76
California Department of Transportation
    (Caltrans), 41, 122–23
California Gold Rush, 32
California High Speed Rail, 68
California Public Utilities Commission,
    141
Cambridge, Mass., 141, 143–44, 191–92, 193
Cambridge Innovation Center, 191–92
Camp, Dave, 255
Canada, 44, 61, 74
Capital Bikeshare, 205
Carbon Disclosure Leadership Index, 42
carbon emissions, 13, 42, 68, 69, 74, 75, 76,
    86, 95, 103, 116, 130, 176
Carbon Nation, 215
Carbon Performance Leadership Index, 42
Cardozo High School, 263
Car2Go, 145, 147, 148, 152, 154
Carlyle Group, 249
Carnegie, Andrew, 33
carpools, 56
cars:
    accidents of, 12, 24, 125, 127, 132–33,
        139, 162, 207, 209, 277
    buyers of, 131–32
    carbon emissions of, 13, 130, 176

communication between, 133–37, 139, 155, 167, 168
of commuters, 9, 14, 140, 176–77
computerization of, 131–32, 133
driverless (self-driving), 7, 11, 119, 137–39, 167, 272
electric, 7, 17, 64, 129–30, 147–48, 157
entrepreneurship in, 140–42, 165
environmental impact of, 13, 130, 147, 148, 176
federal regulations for, 138–39
flying, 272–73
foreign, 128, 129, 138
fuel efficiency of, 9, 149, 157, 167, 178
hybrid, 129
industry for, 34, 50, 120, 122, 128–31, 134, 143
insurance for, 143, 144
license plates for, 152
market for, 143, 147–48, 154, 168
navigation systems for, 151–52
ownership of, 22, 46, 148, 160–61, 171–75, 179, 185, 187
parking for, 22, 140, 145, 148–52, 154, 159–60, 165, 189, 211, 246–47, 270
pollution from, 1, 9–10, 13, 16, 41, 130, 147, 148, 167, 176, 276
population densities and, 34
public attitudes on, 138–39
rentals of, 142, 145–46
safety of, 131, 139, 140, 143, 202
sales of, 128–32
seat belts for, 12
sensors for, 131–33, 138–39
sharing of (ride sharing), 56, 139–48, 151, 152–54, 159–60, 276
smart, 127–39, 159, 160, 276
speed of, 157
technology for, 127–39, 151–52, 160–61
user fees for, 17, 166–67, 268
see also highways and roads
cell phones, 11, 92, 164, 186
see also smartphones
Census Bureau, U.S., 202

Central Loop BRT, 198
Central Maine & Québec Railway, 44
Central Subway Project, 180, 181
Chambers, John, 97
Chan, Mary, 135, 136
Change Masters, The (Kanter), 4
Changi Airport, 112
Charles River, 21
"Charlie on the MTA," 179
Chase, Robin, 23, 143–44
Chatman, Daniel, 177–78
Chetty, Raj, 182
Chevrolet Malibu, 135
Chevrolet Volt, 130
Chicago, 3, 10–11, 13, 24, 28, 29, 47–52, 60, 68–69, 109, 121, 125–26, 145, 172, 174, 183, 185, 194–200, 201, 205–10, 211, 212, 213–15, 218, 245–47, 252–53, 268, 270
Chicago Department of Transportation (DOT), 48–52, 194, 196, 198, 199–200, 205, 206, 208–9
Chicago Infrastructure Trust, 214, 252–53
Chicago Parking Meters LLC (CPM), 246
Chicago Region Environmental and Transportation Efficiency Program (CREATE), 48–52, 68–69, 268
Chicago Skyway, 245–46, 251
Chicago Streets for Cycling, 205
Chicago Transit Authority (CTA), 194–96
China, 22, 27, 28, 32, 58, 63, 66, 161, 202, 243
China Development Bank, 251
Chrysler Corp., 128, 129, 269
"Chunnel" (Channel Tunnel), 11, 227, 229
Cintra, 245
Citibank, 204
Citi Bikes, 204–5, 206
cities, see urban areas
Citigroup, 242
City Year, 43
Civil Aeronautics Administration (CAA), 77, 78

Civil Aeronautics Board (CAB), 77

Civil War, U.S., 52–53, 275

Class I freight railroads, 33, 38, 43, 59, 268

Cleveland, 83–84

Climate Leadership Award, 42

Clinton, Bill, 2, 18, 35, 89, 129, 189, 252, 267

cloud computing, 96–100

Cold War, 14, 80

Coleman, William, 18

Coletta, Carol, 203

"collaborative consumption," 16, 140, 144

collaborative process, 16, 188–90, 275–76

Collaborative Trajectory Options Program (CTOP), 93–94

College to Careers, 215

Commerce Department, U.S., 169, 250, 278

Commercial Club, 106

Commercial Trucks, 146

Common Operational Picture (COP), 49

communications networks, 12, 17, 113, 122, 125, 128, 131, 135–36, 144, 148, 155–67, 192, 215, 272, 277

Community Redevelopment Agency, 230

Community Reinvestment Act (CRA) (1977), 253

community service, 43

commuter railroads, 9–10, 51, 55–57, 60, 68, 70, 199–200, 248

commuting, 9, 14, 140, 176–77, 181–88, 202, 212–13, 218

commuting zones, 182

competitiveness surveys, 16, 178–79, 276

Complete Streets, 206, 212, 216

Confidence: How Winning Streaks and Losing Streaks Begin and End (Kanter), 4

Congestion Mitigation and Air Quality grants, 205

Connecticut, 9–10, 12, 55–57, 62

Conrail, 35, 36, 66

Consumer Electronics Show, 138, 158–59

containerized shipping, 39

containers, 53, 54, 224

Cook, Tim, 133

Copenhagen Wheel, 211–12

Corker, Bob, 57

cost-effectiveness, 30, 40, 68

Cottonwood Pass, 85

Crescent Corridor, 40, 54, 65–66, 68

Crist, Charles, 234

CSR Corp., 63

CSX, 23, 36–37, 39, 41–43, 46, 49, 52–55, 66

Cuba, 231–32

Customs and Border Patrol, U.S., 113

Daimler AG, 147, 148

Daimler Mobility Services, 148, 154

Daley, Richard J., 48, 213–14

Daley, Richard M., 106, 214, 246, 267

Dallas-Fort Worth International Airport, 102, 105, 114

Danielson, Antje, 23, 143–44

Daschle, Thomas, 70

Data Innovation Challenge, 153, 274

Daversa, Joe, 67

Davey, Richard, 125, 154, 180

Deason, Ian, 77, 104

Defense Department, U.S., 33

Delaney, John, 254

Delivery Information Acquisition Devices (DIADs), 161

Delta Air Lines, 79–82, 85, 95, 107–8

Denmark, 202

Denver, 105, 181, 201, 218, 248, 252, 267

Denver International Airport, 105

Denver Union Station Project Authority, 201, 218

Department of Transportation, U.S. (USDOT), 18, 24, 55, 67, 78, 86, 90, 116–17, 121–22, 137, 153, 161, 166, 167, 169, 249, 250, 261, 263, 272, 274

departments of transportation (DOTs), 7, 14, 41, 48–52, 55, 66, 67, 122, 137, 151, 221–41, 243, 262, 270

derailments, 9–10, 27, 44

Detroit, 21, 24, 46, 121–22, 173, 180, 185, 203

DHL, 79

Diaz, Manny, 15, 24, 188, 213, 230–31, 238, 239, 240

diesel fuel, 61

Dietrich, Carl, 272–73

digital billboards, 125–26, 215

District of Columbia Department of Transportation, 55

DiversityInc, 43

Divvy bike stations, 206–10, 212

Dodge Island, 223–24, 227, 240

Doerr, John, 210

Dole, Elizabeth, 200

"dollar vans," 191

Donohue, Thomas, 21, 237

dot-com crash, 18

double-stack operations, 53, 54

Douglas Aircraft, 33–34

Drake, Ann, 3, 49, 224

driverless (self-driving) cars, 7, 11, 119, 137–39, 167, 272

drivers, truck, 39–40

drunk driving, 126

Dubai, 108, 112, 114

Dubai Ports, 249

Dukakis, Michael, 179–80, 181

Duke Energy, 130

Dulles Airport, 180, 201

dynamic pricing, 123–24

Eagle commuter rail project, 181, 248

earthquake detectors, 27

eBay, 143

Econolite Control Products, 123, 126

Economist, 30, 213

education, 19, 159, 184, 214, 215–16, 218, 239, 261, 262–63, 278

Einstein, Albert, 6

Eisenhower, Dwight D., 14, 34, 77

electric automobiles, 7, 17, 64, 129–30, 147–48, 157

electric bicycles, 211–12

electric cars, 7, 17, 64, 129–30, 147–48, 157

Electric Generation, 130

Electronic Data Systems (EDS), 134

elevators, 176

Emanuel, Rahm, 24, 49, 107, 110, 125, 199, 205, 213, 214, 215, 246–47, 252, 253, 267, 270

Embraer, 80

EMD, 63

Emirates airlines, 112–13, 114

Emirates SkyCargo, 114

Empire Builder service, 59

engines, aircraft, 80

enhanced bicycles, 212

ENIAC computers, 97

Enterprise CarShare, 146

Enterprise Holdings, 142, 145–46

Enterprise Rideshare, 146

entrepreneurship, 11–12, 22, 70, 90–91, 140–42, 165, 266, 271–74

environmental impact statements, 55

environmental issues, 13, 43–44, 51, 55, 71, 73, 80, 90, 105, 130–31, 147, 148, 176
    see also air pollution

Environmental Protection Agency, U.S., 42

Etihad Airways, 112–13, 114

European Investment Bank (EIB), 251

European Union (EU), 37, 243, 251

Express Lanes, 124

EZ Pass, 127

FasTracks, 201

Federal-Aid Highway Act (1956), 14, 34, 180–81

Federal Airport Improvement Program, 107

Federal Aviation Act (1958), 78

Federal Aviation Administration (FAA), 24, 74, 78, 84, 85, 86, 87, 90, 92, 93, 94, 95, 98, 103–4, 106, 110, 116, 244, 247, 261, 270, 273

Federal Communications Commission
(FCC), 166, 169, 261
Federal Express, 79, 82, 105, 114, 161, 199
Federal Highway Administration (FHA),
10, 15, 55, 124, 145, 227, 263
Federal Railway Administration, 66
Federal Reserve Bank, 254, 271
Federal Transit Administration (FTA),
181, 196, 198
FedEx Express, 79
Ferre, Maurice, 227
Finch, Bill, 56
FitBit, 211
Flexcar, 143
flight attendants, 91–92, 100
Florida, 221–41, 242, 243
Florida Department of Transportation,
221–41, 243
Florida Supreme Court, 232
flying cars, 272–73
FlyKly, 212
Ford, Gerald R., 18
Ford Focus, 130
Ford Motor Co., 7, 50, 128, 129, 130, 132,
134, 165
Fort Lee, N.J. bridge scandal (2013), 47
Fortune, 43
4G LTE wireless, 135–36, 156, 157, 158,
164, 272
Foxx, Anthony, 153, 240, 250
Foy, Douglas, 176
France, 28, 37
Freeman-Wilson, Karen, 109
freight railroads, 23, 25, 28, 29, 30, 34, 35,
36–55, 59, 62, 69, 71, 162, 189, 224,
247, 268, 269
frequent-flyer programs, 78, 104
fuel efficiency, 9, 74, 75, 80, 81–82,
85–86, 87, 95, 99, 100, 101 149, 157,
167, 178
Fulton Market, 216

Gannett Fleming, 226
Garrett A. Morgan Technology and

Transportation Futures program,
121–22, 262–63
Garvey, Jane, 3, 89, 111, 233–34, 237
Gary, Ind., 109
Gary-Chicago International Airport, 109
gasoline tax, 7, 17, 57, 104, 120, 123, 124,
166, 207, 243, 256, 268
General Catalyst, 144
General Electric (GE), 42, 63, 80, 85, 113,
121, 249
General Electric Transportation, 63, 107
General Motors (GM), 4, 7, 128, 129, 131,
132, 134, 135, 136, 156, 161, 165, 269
General Motors Institute, 3, 131
George, Matt, 191–93
Germany, 202
GE Watt, 157
Gildersleeve, Mark, 98, 99
Girl Scouts, 238, 256
Glaeser, Edward, 176
Global Business Travel Association, 213
Global Connected Consumer, 135
Global Entry kiosks, 113
Global Initiative, 267
Global Innovation Index, 163
Global Technology Index, 163
Goffman, Erving, 209
"Going Under," 239–40
Goldsmith, Stephen, 123
Gomarco, 63
Gomez-Ibanez, Jose, 252
Google, 7, 11, 22, 62, 136, 138, 141, 155, 160,
165, 167, 192, 193, 211
Google Android, 136, 141, 155, 165
Google buses, 193, 211
GPS, 86, 98, 119, 122, 132, 134–35, 147, 159,
161
grade crossings, 29, 48–52, 67, 265
Grange political movement, 33
Grant, Ulysses S., 32
grassroots movements, 20–21, 189–90
Graves, Bill, 39
Gravina Island Bridge, 221
Great Britain, 61

Great Depression, 54, 66

Great Recession, 18, 54, 107, 164, 178, 236

greenhouse gases, 13, 130, 176

Green Line, 179–80, 191, 192

Griffith, Scott, 3, 23, 144–45, 146, 192

gross domestic product (GDP), 74, 243

Grow America, 250

GS15A crossing, 50–52

GSM, 158, 165

Guangzhou, China, 32

Hairston, Anita, 183–84

Harbor Maintenance Trust Fund, 252

Harkin, Tom, 106

Harlem, 205

"Harriet" tunnel-boring machine, 237–41, 256

Harth, Michael, 151

Hartsfield-Jackson International Airport (ATL), 107–9, 267

Harvard Advanced Leadership Fellows, 252, 255

Harvard Business School (HBS), 2–3, 16, 18, 157, 276

Harvard Kennedy School of Government, 4, 151

Haystack app, 148–49

health care, 174, 183

Heartland Corridor, 54, 68

Heathrow Airport, 78, 81, 103

Hertz 24/7 service, 146

Higgins, Brian, 156–58

High Line, 216

high-occupancy toll (HOT) lanes, 123–24

high-speed trains, 1, 6, 27–29, 35, 44, 58–59, 61–68, 70, 103, 200, 273, 275–76

Highstar Capital, 110, 247, 249

highways and roads, 119–69

    accidents on, 125, 127

    communications and, 119, 125, 133–37, 139, 155, 167, 168

    computerization of, 122, 131

    congestion and delays on, 1, 9–10, 15, 16, 24, 25, 39, 40–41, 46, 47–52, 56,

        69, 94, 103, 122–24, 125, 131, 136–37, 138, 140, 150, 153, 157–58, 167, 171–75, 178, 186, 188, 222, 224–25, 277

    congressional oversight of, 128, 131

    data analysis for, 121, 125–27

    defense rationale for, 15–16, 33, 260

    digital billboards on, 125–26, 215

    economic impact of, 123–24, 127

    ecosystem of, 25–26

    environmental impact of, 130–31

    federal funding for, 14, 30, 57, 61, 123–24, 128–29, 161, 169, 242, 243, 261, 268, 275

    federal regulation of, 138–39

    interstate, 14, 22, 34, 40, 120–21, 122, 221

    investment in, 11, 15–16, 119, 136–37, 154

    leadership on, 128, 131, 137, 138–39

    pedestrians and, 126, 174, 175, 176, 187–88, 201, 202, 209, 210–12, 216, 230–31

    potholes in, 10, 120, 277

    public opinion on, 125, 126–27

    public-private partnerships in, 119, 136–37, 154

    railroads compared with, 30, 33–35, 36, 47–52, 56, 67, 68, 69, 75

    repair of, 7, 10, 120, 277

    sensors in, 121, 122–23, 131, 167, 211–12, 265

    signaling on, 125

    smart, 25–26, 119–20, 127, 128, 155–67

    taxation for, 7, 17, 57, 104, 120, 123, 124, 166, 207, 243, 256, 268

    technology for, 119

    toll systems for, 127, 245–46, 270

    traffic lights for, 121–22, 123, 209, 262–63

    transponders for, 127

    truck transportation on, 13, 33, 36, 38, 39–41, 50–52, 53, 62, 69, 161, 165, 197, 222, 224–25, 230–31, 239, 241

    user fees for, 123–24, 243

    see also cars

Highway Trust Fund (HTF), 14, 17, 22, 57, 169, 180–81, 243, 261

Hodgkins, Christopher, 237, 238–39
Hoecker, Marc, 33, 59
Honda Insight, 129
Honeywell International, 80
Howland, Ray, 89, 100
Hughes Network Systems, 134
Hughes Telematics, 159
HumanKiosk, 157
Hunt, Don, 201
Hurricane Katrina, 16
Hurricane Sandy, 142
hybrid cars, 129
Hyperloop, 64–65, 70, 130, 273

IBM, 11, 78, 90, 91–92, 133, 215
ice storms, 10
IDI, 163–64
idle-reducing technology, 42
Immelt, Jeffrey, 12, 23, 58, 113, 164, 249
Incheon Airport, 28
India, 58
Indiana, 242, 245–46
Indiana Toll Road, 245–46
Information Technology and Innovation
    Foundation (ITIF), 124, 164
infrastructure, 1–26, 221–57
    as assets, 243–47
    author's approach to, 1–8, 18, 39, 241–42
    banks for, 6, 26, 251–55, 271
    bottlenecks in, 45–52
    capitalization of, 247–50
    collaborative process for, 16, 188–90,
        275–76
    community relations and, 43, 233–41,
        255–57, 263
    congressional legislation for, 2–3,
        23–24, 252, 254–55, 276
    economic impact of, 6, 18, 21–23,
        221–41, 260–61
    entrepreneurship in, 11–12, 22, 70,
        266, 271–74
    environmental issues in, 13, 43–44, 255
    future trends in, 259–79

    investment in, 3, 4, 13–14, 39, 221–57,
        264–65
    job creation for, 21–22, 237, 259–60
    leadership for, 1–8, 12, 18, 20–24, 39,
        144, 214, 222–23, 250, 254, 255–57,
        266–68, 273–79
    maintenance of, 20–23, 236–37
    markets for, 69
    narrative for, 260–64
    national vs. regional framework for,
        264–69, 272, 275–76
    political aspect of, 23–24, 222–23, 226–
        41, 249–50, 251, 254–57, 266, 274–79
    privatization of, 243–47, 269–70
    public-private partnerships (PPPs)
        for, 24, 26, 31–32, 35, 46–47, 53, 68,
        71, 110–11, 197–98, 214–15, 221–41,
        247–50, 267–71
    public support for, 2–3, 7–8, 20, 277–79
    statistics on, 2–3, 20
    strategic vision for, 20–23
    taxation for, 222, 251, 254, 264–65
    technology for, 11–12, 160–61, 243,
        271–74
    "three R's" (repair, renewal, and rein-
        vention) in, 17, 68, 259–60, 275–76
    vision for, 11–12, 20–24, 160–61, 255–
        57, 265, 268–69, 274–79
    women as leaders for, 3, 188–89, 238
    see also transportation
Infrastructure Investment Summit
    (2014), 250
Infrastructure Ontario, 253
Innovation Centers, 3, 156–58, 191–92
Innovations in American Government
    Awards, 151
"instant bridges," 121
Insurance Institute for Highway Safety, 162
Intelligent Car Coalition, 159
Intelligent Transportation Society of
    America, 122–23, 124
Intelligent Vehicle Highway Systems Act
    (1991), 162

intercity trains, 47, 56, 57–68, 74–75

intermodal connections, 38, 40–41, 62–63, 103, 264

international trade, 222, 224, 261–62

Internet, 7, 11, 15, 82, 92, 97, 135–36, 147, 155, 160, 163, 167, 168

Internet of Things, 97, 155, 163

Interstate 69 Development Partners, 242

Interstate 81, 40

Interstate Commerce Act (1887), 33

Interstate Commerce Commission (ICC), 33, 34

Interstate Highway Act (1956), 122

Interstate Highway System, 14, 22, 34, 120, 122

iPads, 11, 76, 87, 89–93, 99

iPhones, 141, 155–56, 164, 165

Iran, 58

Isolux Infrastructure Netherlands, 242

IT Industry Competitiveness Index, 163

ITS America, 159

Jacksonville, Fla., 43

Japan, 27, 28, 30, 66, 70, 81, 103, 128, 129, 130, 161

Jeffrey Express bus (Jeffrey "Jump"), 196

Jeffrey Parker & Associates, 229

Jeopardy, 133

jet aircraft, 33–34

JetBlue, 77, 82, 95–96, 99, 104, 115

JFK International Airport, 95–96, 101, 102, 114

job creation, 19, 21–22, 40–41, 49–50, 53–54, 55, 71, 74, 106, 107, 108, 109, 162, 237, 259–60

Johns, Robert, 137

Johnson, Andrew, 32

Johnson, Bill, 225, 234, 239, 240

Juneau, Alaska, 85

Kalanick, Travis, 140–41

Kamen, Dean, 210–11, 273

Kansas City Southern Railroad, 38

Karp, Jeff, 151

Katrina, Hurricane, 16

Kawasaki, 63

Keany, Desmond, 88, 98, 99

Kenny, David, 97, 98

Keystone Corridor, 65–68, 70, 273

Keystone Pipeline, 43–44

Khudeira, Soliman, 50

Kiley, Robert, 180

Kingston Trio, 179

KKR, 249

Klein, Gabe, 192, 205–6, 207

Koch, Stephen, 49, 174, 246–47

Koehler, Bryson, 97

Kopelousos, Stephanie, 234, 235

Kopser, Joseph, 153–54, 274

Kotil, Temel, 114–15

Kubly, Scott, 194

Lac-Mégantic, Quebec, 44

LaGuardia Airport, 111

LaHood, Ray, 107

Lamar Advertising, 126

Latinos, 183, 210

Lazowski, Alan, 151

LAZ Parking, 151–52

LED signs, 125

legacy airlines, 80, 83

Lehman Brothers, 233

Lew, Jacob, 250

Lewis, Michael, 10

light rail networks, 22, 46, 104, 188, 238

Lincoln, Abraham, 32, 35, 275

Linimo trains, 28

Little Engine That Could, The, 45, 276

Lockheed Corp., 33–34

locomotives, 27–28, 61, 63

LoDo (Lower Downtown), 201

Logan International Airport, 81, 102, 114, 115, 195

Lohner Electric Chaise, 129

London Underground, 179, 180

long-distance trains, 28, 38, 44–45,
    59–60, 71
Los Angeles, 24, 109, 150, 201
Los Angeles International Airport (LAX),
    109
Louisiana Purchase (1803), 31
Luis Muñoz Marín International Airport,
    110
Lyft, 141, 142–43, 146, 152
Lynskey, Kevin, 228, 234, 235, 236
Lyon, 28

Macquarie, 245–46
maglev (magnetic levitation) trains, 1, 11,
    28, 35, 64, 66, 70
Main Line of Public Works, 65
Malaysian Airlines, 86
Malloy, Dannel, 56
Malone, Ai-Ling, 184–85
manifest destiny, 31
Markey, Edward, 168
Maryland, 200, 248
Maryland Area Regional Commuter Rail
    (MARC), 200
Massachusetts, 4, 20, 23, 62, 125, 154,
    179–80, 183, 189, 191, 192, 193
Massachusetts Bay Transportation
    Authority (MBTA), 20, 60, 62, 179,
    189, 191, 192, 193
Massachusetts Department of
    Transportation (MassDOT), 154
Massachusetts Institute of Technology
    (MIT), 168, 211
Massachusetts Port Authority
    (Massport), 114
mass transit, 1, 10–11, 17, 20–21, 22, 26, 60,
    61, 67, 69, 70, 117, 160, 177–201, 212,
    218–19, 269, 275, 279
MAT Concessionaire, 236–37, 239, 270
Mayer, Marissa, 119–20
McAdam, Lowell, 134, 149–50, 157,
    164–65, 168
McArthur Causeway, 238
McKinsey & Co., 65

Memphis, 82, 105, 114
Menino, Thomas, 213
Mercedes-Benz, 147, 148
Mercedes-Benz Smart Fortwo, 147
Meridiam Infrastructure Fund, 221–22,
    233–34, 235, 236, 237, 250
Metrarail, 60, 199–200
Metro-North Railroad 1548 accident
    (2013), 9–10, 55–57
Metropolitan Atlanta Rapid Transit
    Authority (MARTA), 108, 189–90
Metropolitan Planning Council, 206, 214
Metropolitan Transit Authority (MTA),
    60
Mexican Cession (1848), 31
Meyer, Eric, 148–49
Miami, 176, 188, 190, 213, 227
Miami Access Tunnel (MAT), 3, 20, 23,
    221–41, 242, 248, 249, 251, 252, 256,
    257, 270, 271
Miami Beach, 190, 211, 223–24, 238
Miami-Dade County, 228, 230, 242, 248,
    275
Miami International Airport (MIA), 104,
    110, 226, 228
MIA North Terminal project, 228
Michigan, 7, 122, 137
Michigan, University of, 137, 272
Michigan Department of Transportation,
    7, 137
Mid-Atlantic Infrastructure Exchange,
    267
Midway Airport, 105–6, 110, 213
Mikulski, Barbara, 256
millennials, 174, 272
Miller, Mark, 89, 99
Mills, Karen, 89
Minicucci, Benito, 85–86
minivans, 13
Minneapolis, 10, 21, 124
minorities, 183, 189, 210, 217
Mintle, Theresa, 215
Mission Street renewal, 176
Mitsubishi, 130

"mixed transit," 195

MnPASS transponders, 124

mobile networks, 12, 17, 113, 122, 125, 128, 131, 135–36, 144, 148, 155–67, 192, 215, 272, 277

Mobility Cooperative, 143

monetized assets, 243–47

monopolies, 30, 32–33, 46, 83

Monopoly, 30

monorails, 180

Montero, Judy, 201

Montreal, Maine & Atlantic Railway, 44

Moorman, Wick, 40, 46

Mora, Jorge, 239, 240

Morgan, Garrett A., 121–22, 262–63

Morgan, J. P., 46

Morgan, Tracy, 40

Morgan Stanley, 138

Motive Power, 63

Mulally, Alan, 128–29

MultiShip Service, 161

municipal bonds, 198, 207, 236, 241–42, 253

Murphy, Chris, 57

Musk, Elon, 64–65, 70, 130, 273

National Aeronautics and Space Administration (NASA), 77, 98, 168

National Environmental Policy Act (1969), 51

National Express Group, 110

National Gateway, 52–55, 68

National Highway Traffic Safety Administration, 12, 127

National Household Travel Survey, 202–3

National Interstate and Defense Highways Act (1956), 14, 34, 180–81 (1972)

nationalization, 34, 35, 269

National Railroad Passenger Corp., *see* Amtrak

National Surface Transportation Policy and Review Study Commission, 243

National Transportation Safety Board (NTSB), 57, 167

National Weather Service, 97, 277–78

Native Americans, 32

natural gas, 42

navigation technology, 25, 83–87, 92, 151–52

Netherlands, 202, 203–4

Network Fleet, 157

New Car Shipping Center, 50

New Jersey Department of Transportation (NJ-DOT), 125

New Orleans, 29

New York City, 9, 10, 56, 60, 141, 142, 149, 179, 180, 183, 191, 204–5, 206, 213, 216, 269–70

New York State, 127, 142

New York Times, 40

NextGen technology, 76, 84, 86, 87, 88, 89, 93–94, 100–101, 116–17, 161, 166, 243, 271–72

Nicholson, Pamela, 146

Nigeria, 12, 164

Nissan Leaf, 130

Nixon, Richard M., 35

Niznik, Tim, 93

noise reduction, 76, 86, 87, 103

Nolan, Robert, 177–78

Norfolk Southern, 36, 38, 40, 46, 51, 54, 65, 66, 69

North American Free Trade Agreement (NAFTA), 38

North American Vehicle Operations, 134

Northeast Corridor, 10, 29, 33, 58–60, 62, 70, 74, 266

Northeast Maglev, 70

Northeast Regional route, 58–59

Northrop Grumman Corp., 80

Norway, 211

Obama, Barack, 18, 214, 250

Occupy Crosswalk, 187

O'Hare International Airport (ORD), 102, 105–6, 109, 114, 213, 267

Ohio, 54

oil pipelines, 43–44, 45

Oliver, Persephone, 126

O'Malley, Martin, 267

Omidyar, Pierre, 143

O'Neill, Thomas P. ("Tip"), 180–81, 266, 275

One World Alliance, 111

OnStar system, 134–35, 159, 161

On the Fly, 205

Open Automotive Alliance, 136, 165

Open Skies agreements, 18, 111–12, 113, 264

Operation '305,' 237

optimized profile descents, 87

Oregon, 7, 268

Organization for Economic Cooperation and Development (OECD), 46, 57, 61

overhead clearance, 52–53

Pacific Bell, 134

Pacific Railroad Acts (1861), 32

Palin, Sarah, 221

Panama Canal, 46, 54, 69, 222, 225

Pan American Airways, 77

Paris, 28

park-and-ride lots, 189

Parker, Jeffrey, 233

Parker app, 149

parking, 22, 140, 145, 148–52, 154, 159–60, 165, 189, 211, 246–47, 270

parking apps, 22, 148–49, 165

parking meters, 151, 246–47, 270

Parking Panda, 150

parking rates, 246–47

ParkMe, 165

ParkMobile, 150

Partnership for a New Generation of Vehicles, 129

Partnership to Build America Act, 254–55

passengers service, 73, 74, 77–78, 93–94, 99–100, 102

Pathways in Technology Early College High School (P-TECH), 216, 263

Patrick, Deval, 23, 154

pedestrians, 26, 126, 157, 174, 175, 176,
187–88, 201, 202, 209, 210–12, 216, 230–31

Pego, Gus, 235

Pennsylvania, 66, 67, 183, 247, 273

Pennsylvania Central Railroad (Penn Railroad), 35, 54, 65–66

Pennsylvania Department of Transportation (PennDOT), 66, 67

Pennsylvania Station, 33

Pennsylvania Turnpike, 247

pension funds, 264–65

People Express, 82

People Mover, 180

Pepper, Claude, 227

Philadelphia, 60, 183

pilots, 75–76, 85, 87, 88, 90–94, 95, 100, 116, 273, 277–78

Piper Jaffray, 253

Pittsburgh, 104–5

placemaking, 175–76

population density, 34, 69, 161, 171, 172, 173–74, 190

Porsche, Ferdinand, 129

Porsche AG, 108

Port Authority of New York and New Jersey, 110–11, 267

Porter, James, 196

Portland, Ore., 203, 278

Port of Baltimore, 53

Port of Miami, 20, 23, 221–41, 242, 248, 249, 251, 252, 256, 257, 270, 271

Postal Service, U.S., 79

potholes, 10, 120, 277

Potts, William, 121, 124

poverty, 26

Powerful Answers grants, 159

Preqin, 250

President's Council on Jobs and Competitiveness, 23, 164

Prime Time, 142–43

Pritzker, Penny, 250

private activity bonds, 242

privatization, 109–11, 243–47, 269–70

Privett, Keith, 197–98, 208

Project Ginger, 210

property taxes, 275

Providence, R.I., 62

public-private partnerships (PPPs), 24,
26, 31–32, 35, 46–47, 53, 68, 71, 110–
11, 197–98, 214–15, 221–41, 247–50,
267–71

public transportation, 1, 10–11, 17, 20–21,
22, 26, 60, 61, 67, 69, 70, 117, 160, 177–
201, 212, 218–19, 269, 275, 279

Public Works Administration (WPA),
54

Puentes, Robert, 59

Puerto Rico, 247, 271

Pullman Rail Journeys, 29

Purple Line, 248

Qantas Airlines, 111

Qatar Airways, 112–13

Queens, 205

radar, 132

radio navigation, 83–84, 85, 86, 92

Rail Passenger Service Act (1970), 58

railroads, 27–71
accidents in, 9–10, 27, 37, 44, 55–57
agriculture and, 33
airline industry compared with,
34–35, 59, 117, 161–62, 178
antiderailment devices for, 27
bankruptcies in, 34, 66
bottlenecks in, 47–52, 53
bullet trains for, 27, 35, 64, 66, 70, 103
bureaucracy of, 52, 60, 67
capitalization of, 35, 38, 41, 42, 61–62
carbon emissions of, 13, 42, 68, 69
Class I freight railroads in, 33, 38, 43,
59, 268
Class II freight railroads in, 38
Class III freight railroads in, 38
commuter, 9–10, 51, 55–57, 60, 68, 70,
199–200, 248
competition in, 30, 32–33, 46–47, 54,
59, 60–64, 69

congestion and delays in, 9–10, 29,
45–52
congressional oversight of, 32, 44–45,
55, 61, 71
consolidation of, 30, 32–33, 46, 48
containers carried by, 53, 54, 224
cost-effectiveness of, 30, 40, 68
customer satisfaction in, 35, 62, 64, 69
derailments in, 9–10, 27, 44
economic impact of, 31, 38–39, 71
efficiency of, 38–41, 53, 61
electrification of, 27–28, 29, 65, 67
environmental impact of, 43–44, 51,
55, 71
European, 30, 37, 62, 66
federal funding for, 31–35, 46, 52, 61
freight, 23, 25, 28, 29, 30, 34, 35, 36–55,
59, 62, 69, 71, 162, 189, 224, 247, 268,
269
fuel efficiency of, 40, 42, 53
grade crossings in, 29, 48–52, 67, 265
high-speed trains in, 1, 6, 27–29, 35,
44, 58–59, 61–68, 70, 103, 200, 273,
275–76
highways and roads compared with, 30,
33–35, 36, 47–52, 56, 67, 68, 69, 75
hyperloop system for, 64–65, 70, 130,
273
intercity, 47, 56, 57–68, 74–75
intermodal connections for, 38, 40–41,
62–63, 264
international standards of, 58, 61, 62,
63–64
investment in, 38–39, 41, 48–55, 57,
60–62, 64, 70–71, 172, 264
job creation and, 40–41, 49–50, 53–54,
55, 71
labor sources for, 32, 33
leadership in, 22, 31, 33, 34–35, 64,
68–71
light rail networks in, 22, 46, 104, 188,
238
local and regional initiatives for,
48–52

railroads (*continued*)

locomotives for, 27–28, 61, 63

long-distance, 28, 38, 44–45, 59–60, 71

maglev (magnetic levitation) trains for, 1, 11, 28, 35, 64, 66, 70

management of, 35, 36–37, 41–43, 60–61, 65

modernization of, 48–55

monopolies in, 30, 32–33, 46, 83

nationalization of, 34, 35, 269

oil transported by, 43–44, 45

outdated equipment for, 29, 30–31, 48–52

overhead clearance for, 52–53

passenger, 1, 25–30, 34, 35, 44–45, 46, 47, 48–49, 51, 57–68, 71, 199–200, 269

political aspect of, 22, 31, 33, 34–35

profitability of, 36–37, 38, 49–60, 67

public opinion on, 30, 33, 68–69, 71

public-private partnerships (PPP) for, 31–32, 35, 46–47, 53

railcars in, 63

rates charged by, 33, 36–37

real estate of, 31, 32

reinvestment in, 38–39, 41

repairs for, 6, 68–69, 70, 160

revenue losses of, 59–60

routes for, 28–29

safety performance of, 42–43

signaling and switching systems for, 28, 29, 49, 67, 69

single-track, 53

taxation for, 49, 57

technology for, 28, 48–55, 66, 160

terminals for, 199–200

TGV (Train à Grande Vitesse) trains for, 28, 66

tickets for, 70–71

tracks for, 9–10, 38, 41, 46, 66, 70

train manufacturers for, 63

transcontinental, 22, 31, 33, 35, 266, 275

tunnels for, 52–55

urban, 24, 46, 177–78, 180, 186, 187, 190–91, 195, 217–18, 269, 279

Ramirez, Jorge, 253

Random Acts of Transit, 187–88

Reagan National Airport, 200

real-time data, 125, 157

Reed, Kasim, 108–9

Regalado, Tomas, 240

Reilly, Sean, 126

Related Companies, 204

remote work, 119–20

Rendell, Edward, 247

request for proposal (RFP), 234

required navigation performance (RNP), 85–86

REQX, 204–5

Research and Innovative Technology Administration, 162

Rhode Island, 10, 189

Rhode Island Public Transit Authority (RIPTA), 189

Rice, Tom, 254, 255

RideScout, 151–54, 165, 274

ride sharing, 56, 139–48, 151, 152–54, 159–60, 276

roads, *see* highways and roads

Rockefeller, John D., 33

Rockefeller, John D., IV, 44–45

Rockefeller Foundation, 198

Rometty, Virginia ("Ginni"), 3, 12, 133

Roosevelt, Franklin D., 77

Rose Fitzgerald Kennedy Greenway, 176

runways, 101, 105–7, 108, 109

Rus, Daniela, 3, 168

Russia, 37, 58

SABRE system, 78

St. Paul, Minn., 124

sales taxes, 188–89

Salt Lake City, 146

Salvucci, Fred, 180, 181

Sampson, Robert, 183

S&P 500, 42

Sandy, Hurricane, 142

San Francisco, 122, 141, 142, 143, 149, 151, 176, 179, 180, 182–83, 192, 193

San Francisco Bay Bridge, 122

San Francisco Department of Transportation, 151

San Jose, Calif., 182, 203

San Juan, Puerto Rico, 271

Sargent, Francis, 180–81

SAS, 78

satellite navigation, 84–87

school districts, 159

science, technology, engineering, and math (STEM) schools, 215–16, 239, 262–63

Scott, Beverly, 3, 188–89, 190, 266

Scott, Rick, 240

Seattle-Tacoma Airport, 102, 105

Second Avenue line, 180

Segway, 210–11, 273

Seidenberg, Ivan, 157, 158–59

Sen, Laura, 224

Senate, U.S., 6, 44–45
    Committee on Commerce, Science, and Transportation of, 44–45

SENSEable City Lab, 211

sensors, 121, 122–23, 131, 167, 211–12, 265

Sensys Networks, 122–23

Seoul, 28

September 11th attacks (2001), 18, 81, 104–5

sewer lines, 172

SFPark, 151

"shadow tolls," 229

"shadow vans," 192

Shanghai, 1, 11, 28

Shanghai Pudong Airport, 11, 28

Shanghai Transrapid, 28

"sharing economy," 16, 140, 143, 175

Shinkansen trains, 27, 70

SideCar, 141, 143, 152

Siemens, 107

signaling and switching systems, 28, 29, 49, 67, 69

Silicon Valley, 132, 133, 192

"silos," 19–20, 117, 189, 262

Silver Line, 180, 191, 195

Silvers, Damon, 252–53

Singapore, 168

Singapore Airlines, 112

"sin products," 104

Sky Team Cargo, 80

Skytrax, 102

Slater, Rodney, 2, 18, 121–22, 129, 189, 253, 262–63

Small Business Administration, U.S., 169

Small Business Innovation Research grants, 169, 274

Smart Automobile, 130, 148

smart cards, 127

smart cars, 127–39, 159, 160, 276

Smarter Cities Challenge Summit, 4

"smarter planet," 11

smarter trash barrels, 157–58

smartphones, 7, 76, 91, 92, 119, 120, 131, 133, 136, 137–38, 141, 155–56, 157, 164, 165, 276

smart pill bottles, 158

smart roads, 25–26, 119–20, 127, 128, 155–67

Smith, Fred, 79

solar power, 64–65

Southeastern Pennsylvania Transportation Authority (SEPTA), 60, 65

South Korea, 28

Southwest Airlines, 82

Soviet Union, 14, 22, 77

Sputnik launching (1957), 14, 22, 77

Sriver, Jeffrey, 50, 208

Staggers Act (1980), 34

State Department, U.S., 112

Stein, Gertrude, 123

Stewart International Airport, 110

Storch, Gerald, 224

Street Bump, 121

Streetline, 150

Streetsblog Chicago, 196

Stutler, Denver, 228

subways, 177–78, 180, 186, 187, 190–91, 195, 217–18, 269, 279

    see also specific systems

Superpedestrian, 211–12

surge pricing, 142

SUVs, 13

Synovia Solutions, 159

Systematic, Project Expediting, Environmental Decision-making (SPEED), 51

System Operations Control Center, 75–76

Tappan Zee Bridge, 24

TAPS software, 98

tarmac delay penalties, 95–96

taxation, 7, 14, 75, 103–4, 111, 188–89, 197, 198, 207, 222, 251, 254, 264–65

taxi industry, 22, 101, 120, 140–42, 168, 276

Tax Increment Financing bonds, 198, 207

Taylor, Andy, 146

Teletype, 88

Temasek, 112

Terrafugia, 272–73

Tesla Model S, 130

Tesla Motors, 64, 130, 273

Tesla Roadster, 130

Tesla Space X, 273

Texas Central Railway, 70

Texas Eagle service, 59

texting, 88, 92

T. F. Green Airport, 62

TGV (Train à Grande Vitesse) trains, 28, 66

The Weather Company (TWC), 3, 97–98, 273, 277–78

Third World, 12, 51, 164

31L/13R runway, 101

3G wireless networks, 158, 161

3M, 152

"three R's" (repair, renewal, and reinvention), 17, 68, 259–60, 275–76

ticket prices, 70–75, 78, 82, 103–4

Time, 144

Tokyo, 103

toll systems, 127, 245–46, 270

Tolva, John, 215

ton-miles, 41

Toronto, 122

Total Turbulence, 98, 100

Toyota, 128, 129

Toyota G21 project, 129

Toyota Prius, 129

tracks, railroad, 9–10, 38, 41, 46, 66, 70

traffic lights, 121–22, 123, 209, 262–63

Transcontinental Railroad, 22, 31, 32, 35, 266, 275

Transit Action Committee (TrAC), 187–88

transit-oriented development, 31

Transit Score, 182

transportation:

    accidents of, 9–10, 29

    bridges for, 1, 7, 10, 15, 21, 23, 50, 51, 52, 120, 122, 172, 221, 252, 256, 264, 265

    bus, 10–11, 21, 22, 26, 61, 117, 157, 159, 172, 177–78, 184, 185, 186–87, 190–201, 206, 207, 208, 209, 211, 215, 218, 268, 273

    business interests and, 18–19, 20, 276

    carbon emissions in, 12

    collaboration in, 7, 197–98

    commuter, 9, 14, 140, 176–77, 181–88, 202, 212–13, 218

    congestion in, 9–10, 29

    congressional legislation for, 5–6, 14, 18, 20, 180–81, 213, 262–63

    consumer movements for, 20–21

    corridors of, 10, 29, 33, 40, 41, 54, 58–60, 62, 65–68, 70, 74, 266, 273

    data analysis for, 24, 96–97, 119, 121–27, 128, 140, 155, 157, 166, 167, 192, 262, 263, 273 delays and breakdowns in, 9–10, 11, 29, 47–48

economic impact of, 19, 21–22, 26

environmental issues and, 16, 19, 167–68

European, 202, 203–4, 209, 243, 250, 251

for families, 15, 19, 202–3

fares for, 62, 194–95

federal funding for, 180–81, 214, 223, 241, 265, 275–76

federal regulation of, 24, 161–62, 167–68

foreign suppliers for, 20, 229–32, 242

government spending for, 5–6, 18, 19–20

infrastructure of, 1–8, 166–67

intermodal connections for, 38, 40–41, 62–63, 103, 264

international trends in, 4

investment in, 3, 14–15, 21, 71, 180, 218–19

job creation in, 19, 162

leadership in, 12, 18, 24–26, 167–69, 259–60, 274–79

local and regional, 4, 6–7, 19–20, 26, 46–47

"mixed transit" in, 195

mobility challenge in, 15, 24–26, 181–88, 260–64, 278–79

political aspect of, 5, 188–90

priorities in, 5, 263–64

private-public partnerships (PPPs) in, 24, 26, 71, 110–11, 197–98, 214–15

public (mass transit), 1, 10–11, 17, 20–21, 22, 26, 60, 61, 67, 69, 70, 117, 160, 177–201, 212, 218–19, 269, 275, 279

public opinion on, 20

safety in, 202, 278

"silos" for, 19–20, 117, 189, 262

smart, 160–69, 262

social impact of, 19, 181–88

speed of, 195–96, 197

strategic vision for, 2, 5, 7–8, 21–22

taxation for, 7, 14, 188–89, 197, 198, 207

technology for, 9–26, 160–69

ticketing in, 70–71

transfer points in, 199–201

in urban areas, 1, 26, 61, 177–201

see also specific types of transportation

Transportation Equity Caucus, 187

Transportation Infrastructure Finance and Innovation Act (TIFIA) (1998), 236, 249, 252

Transportation Investment Generating Economic Recovery (TIGER), 54

Transportation Research Institute, 137

Transportation Security Administration (TSA), 81

TransTech Academy, 263

Treasury bonds, 57

Treasury Department, U.S., 57, 129, 249, 250

trucking industry, 13, 33, 36, 38, 39–41, 50–52, 53, 62, 69, 161, 165, 197, 222, 224–25, 230–31, 239, 241

Truman, Harry S., 77

Trumka, Richard, 21, 237

truss bridges, 52

Tubman, Harriet, 238

Tuchman, Maxeme, 222, 238

tunnel-boring machines, 237–41, 256

tunnels, 3, 52–55, 221–41, 242, 256

see also Miami Access Tunnel (MAT)

Turkish Airlines, 114–15

Uber, 22, 140–42, 147, 153, 154, 165, 192, 263, 273

Union Pacific Railroad, 35, 38

Union Station (Chicago), 199–200, 218

Union Station (Denver), 201, 218, 252, 267

Union Station (Washington, D.C.), 200–201, 218

United Airlines, 95, 99, 106–7

United Arab Emirates (UAR), 108, 112–13, 114

United Nations International
Telecommunication Union, 163–
64, 262
United Parcel Service (UPS), 79, 161
United Streetcar, 63
United Technologies, 80
urban areas, 171–219
assets of, 243–47
bicycles in, 175, 176, 186, 201, 202–12,
218, 265, 269–70, 276, 279
buses in, 177–78, 184, 185, 186–87,
190–201, 206, 207, 208, 209, 211,
215, 218, 273
businesses in, 175, 197–98, 215
carbon emissions in, 176
cars in, 178, 185–86, 187, 195, 198,
202, 207, 209, 210
commuting in, 181–88, 202, 212–13,
218
cultural institutions of, 225–26,
230–31
economy of, 175, 178, 181–88, 214–15,
217
education in, 184, 214, 215–16, 218, 239
entrepreneurship in, 175, 191–93,
210–12
environmental issues in, 176, 213
health care in, 174, 183
investment in, 178, 218–19
leadership in, 177, 212–19
minorities in, 183, 189, 210, 217
parking in, 176, 209, 211, 270
pedestrians in, 26, 126, 157, 174, 175,
176, 187–88, 201, 202, 209, 210–12,
216, 230–31
placemaking in, 175–76
political situation in, 188–90, 206–7
pollution in, 171
population of, 34, 69, 161, 171, 172,
173–74, 190
property values in, 178
renewal of, 174–75, 176

safety in, 202, 207, 209
schools in, 175
social mobility in, 181–88
streets in, 171–72, 174, 177, 190–91,
206, 212, 216–19, 264
suburbia compared with, 173, 174–77,
217, 263–64
subways in, 177–78, 180, 186, 187,
190–91, 195, 217–18, 269, 279; *see
also specific systems*
taxation in, 188–89, 207
unemployment in, 183, 185, 189, 194,
217
US Airways, 83
U.S. Competitiveness Project, 2
user fees, 17, 166–67, 268
"Use Yah Blinkah" campaign, 125
US Railcar, 63

Vaisala, 123
Vanderbilt, William, 33
vehicle charging stations, 130
vehicle-miles-traveled fees, 7, 17, 166, 268
vehicle-to-infrastructure (V2I) com-
munication, 134, 139
vehicle-to-vehicle (V2V) communica-
tion, 133–37, 139, 155, 167, 168
Verizon Communications, 150, 155–60,
161, 165–66, 168, 191–92, 272
VGO telepresence robots, 119, 158
Viciedo, Marta, 187–88
video cameras, 132
VINCI Park, 151
Virginia Avenue Tunnel, 52–55, 273
Virginia Railway Express (VRE), 200
visas, 113–14
Volkswagen, 128
Volpe National Transportation Systems
Center, 137, 272

Walker, David, 249
Walk Score, 182

Walmart, 40

War Bonds, 256

Ward, Kim, 43

Ward, Michael, 23, 36–37, 39, 41, 43, 46, 53

Warner, Waide, 255

Washington, D.C., 21, 52–55, 124, 144–45, 179, 180, 191, 200–201, 211, 218, 263

Washington Beltway, 124, 200

Washington Metropolitan Area Transit Authority, 263

Washington Metrorail, 179, 180, 200–201

Washington State, 124

Watson supercomputer, 133

weather conditions, 10, 25, 73–74, 75, 76, 85, 87–88, 91, 92, 94–96, 99–100, 105–6, 277–78

Weather Services International (WSI), 98

Weigert, Karen, 215

West Coast Infrastructure Exchange, 266–67

White Bicycle Plan, 203–4

Wiedel, Sean, 198, 208

Wi-Fi networks, 12, 17, 113, 122, 125, 128, 131, 135–36, 144, 148, 155–67, 192, 215, 272, 277

Will, Brian, 23, 75–76, 83, 86, 88, 89, 91, 100, 116

WiMAX, 156

Wings Club, 77

Winsten, Jay, 126

Woods Hybrid, 129

World Business Chicago, 214–15

World Class: Thriving Locally in the Global Economy (Kanter), 4

World Economic Forum, 74, 113, 210

World Economic Forum Global Competitiveness Report (2013), 74

World's Most Admired Companies, 43

World War II, 14, 30, 33, 34, 35, 80, 128, 256

Wright, Orville and Wilbur, 77

WSI Forecaster system, 98

WSI Fusion system, 98

Zakim Bunker Hill Bridge, 230

Zimride, 142, 146

Zipcar, 3, 22, 23, 139–40, 143–48, 152, 165, 205, 263, 273, 276

ABOUT THE AUTHOR

Rosabeth Moss Kanter holds the Ernest L. Arbuckle Professorship at Harvard Business School, where she specializes in strategy, innovation, and leadership for change. She is also Chair and Director of the Harvard University Advanced Leadership Initiative, which helps successful leaders at the top of their professions apply their skills to national and global challenges in their next life stage. The former chief editor of *Harvard Business Review*, Professor Kanter has been named one of the "50 most powerful women in the world" (*Times of London*), and the "50 most influential business thinkers in the world" (Thinkers 50). She is the best-selling author or coauthor of nineteen books, including *The Change Masters, Confidence: How Winning Streaks and Losing Streaks Begin and End*, and *Men and Women of the Corporation*, and is also a prominent advisor to CEOs and senior executives and serves on numerous boards and advisory councils. She is a long-standing trustee of City Year, a model for civilian national service, which serves at-risk students in urban schools in twenty-six U.S. cities and in the UK and South Africa.